CW00727505

HOUSE *of* KARLS

Also by Dr Karl Kruszelnicki

Curious & Curiouser

Brain Food

50 Shades of Grey Matter

Game of Knowns

Dinosaurs Aren't Dead

Dr Karl's Big Book of Science Stuff and Nonsense

Dr Karl's Even Bigger Book of Science Stuff and Nonsense

DR KARL KRUSZELNICKI

HOUSE of KARLS

First published 2014 in Macmillan by Pan Macmillan Australia Pty Ltd

1 Market Street, Sydney, New South Wales, Australia, 2000

Cataloguing-in-Publication entry is available

from the National Library of Australia

http://catalogue.nla.gov.au

Cover, internal design and typeset by Xou Creative, www.xou.com.au

Cover photographs of Dr Karl by Mel Koutchavlis

Internal illustrations by Douglas Holgate

Printed by McPherson's Printing Group

I dedicate this book to those scientists who suffer for their work. Not since Galileo has there been such a well-organised campaign against people searching for the truth.

Einstein (relativity) didn't get orchestrated hate mail. Fleming (penicillin) didn't receive suspicious packages in the post. Newton (maths, calculus, gravity, etc) didn't experience repeated death threats.

But this is happening right now. To all those scientists: keep calm, stick to the facts and the tide will turn (even while it's rising).

Let me toast the Climate Scientists who are taking the heat for all of us.

CONTENTS

01

60 PLANETS TO LIVE ON

For me, one major problem with Space Travel is that there are no relatives or friends to visit who happen to be living on nearby habitable Earth-like planets. That's because there are no other habitable Earth-like planets in our Solar System.

Venus is out, because it's too hot (thanks to a runaway Greenhouse Effect). We can exclude Mars also, because it's too cold (not enough atmosphere to give a mild Greenhouse Effect). I do love the International Space Station. While it does have a comfortable shorts-and-T-shirt environment on the inside, you have to wear lots more than a T-shirt when you go outside.

GOLDILOCKS ZONE

Earth is located in a prime real estate location – the so-called "Goldilocks Zone".

If you ever read the fairytale of *Goldilocks and the Three Bears*, you will remember that Goldilocks found that one bowl of porridge was too hot, one was too cold, but one was just right. So the astronomers (who all read their fairytales as children) describe our planet Earth as being in the Goldilocks Zone.

Inside the Goldilocks Zone, water can exist as liquid water, not just as freezing ice or scalding steam. Earth is not too hot (like Venus) and it's not too cold (like Mars) – it's just right.

My ideal Solar System would have a whole bunch of habitable Earth-like planets. Of course, to be habitable, they would have to be inside the Goldilocks Zone. This would have the added advantage that I wouldn't have to spend too much time travelling to visit family and friends on other planets.

FANTASY STAR SYSTEM

Of course, this would mean shifting planets around the Solar System! (This is a Big Ask, considering we currently can't even shift Mount Everest.)

We have to assume that the human race of the future will be able to control massive energies. Right now this is just science fiction, but looking thousands of years into the future, who knows?

Sean Raymond, an astrophysicist from Bordeaux Observatory in France, made this assumption for his game called "Fantasy Solar System".

Because he's a physicist, his Fantasy Solar System with 60 habitable Earth-like planets had to follow the Laws of Physics. More specifically, his arrangement of planets had to be both scientifically plausible in the short term, and gravitationally stable over the long term – that's billions of years. After all, he wanted life to be able to evolve.

He tried a few different options on the way to his final design.

More Than One Planet Per Orbit? Yes – Lagrangian Points

You can have more than one planet in a single orbit around a star. The person who worked this out was the mathematician Joseph-Louis Lagrange. He did this in 1772 in his "Essay on the Three-Body Problem".

If you consider the simple case of the Earth orbiting around the Sun, there are five so-called "Lagrangian Points". These are the locations where you can insert a third body – and have it stay there, in a stable orbit. (This assumes there are no other planets in the solar system, and that the third body is smaller than the Earth.)

The first three Lagrangian Points are on the straight line extending through the Earth and the Sun. Lagrangian Point 1 (L1) is between the Earth and Sun – about 1.5 million kilometres from the Earth. Lagrangian Point 2 (L2), also about 1.5 million kilometres from the Earth, is on the side of the Earth away from the Sun. L1 and L2 are not in Earth's orbit.

Lagrangian Point 3 (L3) is on the other side of the Sun (on a line extending through the Earth and the Sun), and in the Earth's orbit.

Lagrangian Points 4 and 5 (L4 and L5) are not on this imaginary line. L4 is in the Earth's orbit, and 60° ahead of the Earth. L5 is also in the Earth's orbit, but 60° behind it.

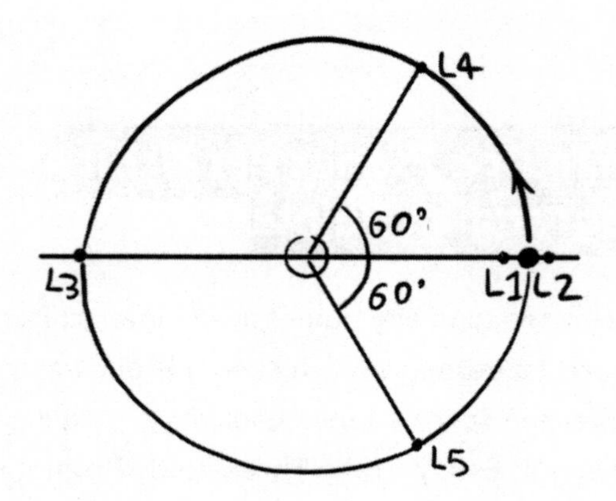

ULTIMATE SOLAR SYSTEM NUMBER 1

For his Ultimate Solar System Number 1, Sean Raymond chose not a star like our Sun, but a smaller one – a Red Dwarf. Because a Red Dwarf has a lower mass, it would last a lot longer than our star, giving life on our planet longer to evolve and survive. Of course, the star wouldn't be as bright, but you just shift your planets in a bit closer to the Red Dwarf. (Don't worry about how you shift them, that's a "minor" Engineering Problem. Let's just stick to the Physics.)

Sean Raymond then worked out how to have 24 habitable Earth-like planets orbiting this single Red Dwarf star.

The first trick is to realise that you can have two planets in the same orbit around a star. You just have to locate them 60° apart, and make sure they have roughly similar masses. Yup, two planets in the same orbit can be perfectly stable.

The second trick is to realise that one Earth-sized planet can have an almost Earth-sized body dancing around it. This is similar to the Earth–Moon system, but in this pair the Moon would be almost as big as the Earth. (To be more accurate, these locked planets would circle their common centre of gravity.) So in an orbit you get two pairs of Earth–Earth systems, or four planets per orbit.

The third trick is to know that the Goldilocks Zone is wide enough to carry six of these orbits. That's 6 × 4 = 24 planets. Sure, the conditions would be warmer for the planets in the orbits closer to the Sun – and cooler for the orbits further from the Sun. But water would still be a liquid, and life could exist. The Laws of Physics tell us that these orbits would be stable over billions of years.

So, to recapitulate, we began with two Earth-like planets dancing around each other, and then added another pair in the same orbit. This gave us four Earths in a single orbit. It turned out we could have six such orbits around a Red Dwarf, which gave us 24 Earths. Suddenly we have created Ultimate Solar System Number 1, with 24 potentially habitable planets.

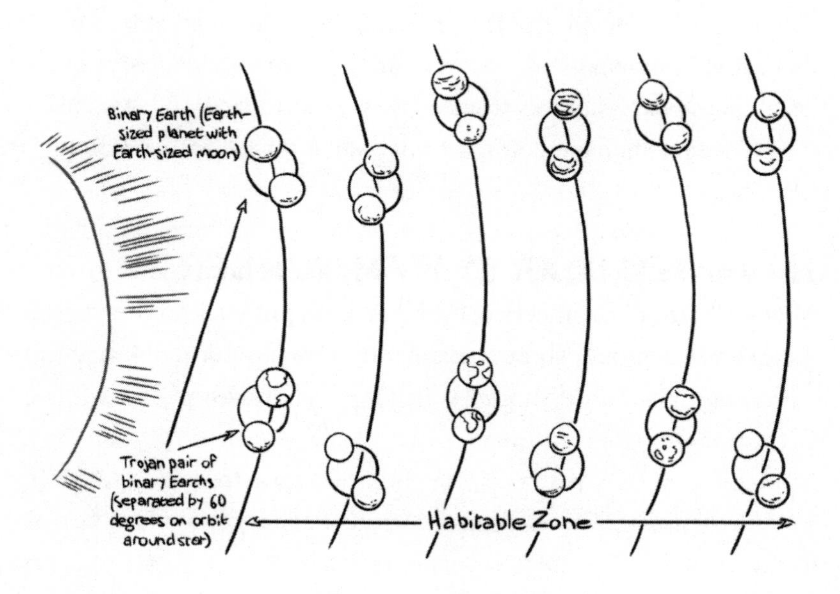

Satellites and Rocks in Space

Lagrangian Point 1 (L1) is quite a popular orbit for spacecraft. It's between the Earth and the Sun. The international *Sun-Earth Explorer 3* satellite spent four years there – from 1978 to 1982. The *Solar and Heliospheric Observatory* (*SOHO*) and the *Advanced Composition Explorer* (*ACE*) arrived at L1 in 1996 and 1997, and are still there.

Several spacecraft (*WMAP, PLANCK, GAIA* etc.) have been sent to L2, on the far side of the Earth away from the Sun.

In October 2010, we discovered a 300 metre asteroid in the L4 Point. It's called 2010 TK_7.

We have discovered about 6000 asteroids in the L4 and L5 Points of Jupiter – 60° ahead of it, and 60° behind it, compared to Jupiter's orbit. They range in size from 200 kilometres in diameter and down. But there are many more that we have not yet found. We estimate, from deep surveys of limited areas of the sky, that there are probably over a million asteroids bigger than one kilometre across in the L4 and L5 Points.

ULTIMATE SOLAR SYSTEM NUMBER 2

Then Sean Raymond let his astrophysicist imagination go crazy. He worked out a different way to stuff more Earth-like planets into a Solar System.

In our Solar System, the giant planet Jupiter has some quite large moons orbiting it. Why not substitute Earths? Sean Raymond began with a single Jupiter-sized planet in an orbit around a star – in the Goldilocks Zone. He then asked himself, how many Earth-sized planets could orbit this Jupiter?

The answer turned out to be five. Even better, the orbits around this Jupiter would be stable for billions of years.

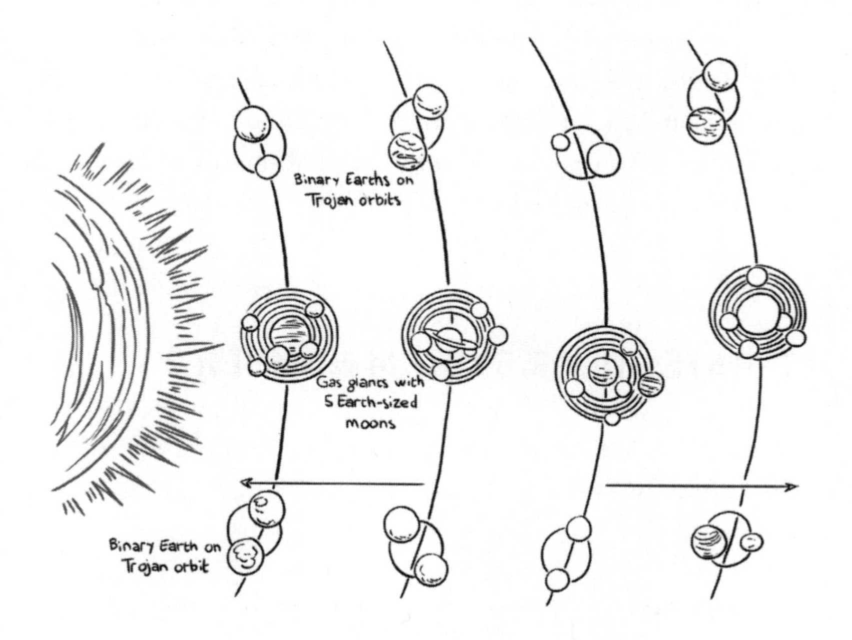

And then, using our new Best Friend Forever, Lagrangian Points, he realised that he could stuff in a few more Earth-sized planets. One would be leading 60° ahead of the Jupiter-with-five-orbiting-Earths (L4), and the other would be trailing 60° behind it (L5) in the same orbit.

And when he did the Physics, it turned out that it would all be gravitationally stable.

Then Sean Raymond went a little further. He changed the leading and trailing Earths into double-Earths (as in our Ultimate Solar System Number 1, similar to Earth–Moon systems, but with the Moons closer in size to the Earth). He again checked the Physics – and that configuration would be stable for billions of years.

That gave a total of nine Earths in one Solar orbit – a Jupiter-with-five-orbiting-Earths, two Earths circling each other at 60° in front of the Jupiter and another two Earths circling each other at 60° behind the Jupiter.

But could he stuff six orbits inside the Goldilocks Zone? Unfortunately, no – because Jupiters are big, and need lots of room.

But he could shove four orbits into the Goldilocks Zone. Four orbits multiplied by nine Earths per orbit works out to 36 Earth-like planets around a single Red Dwarf. That was Ultimate Solar System Number 2.

ULTIMATE SOLAR SYSTEM NUMBER 3

Which one should Sean Raymond go for: 24 planets or 36?

The answer was obvious. Both.

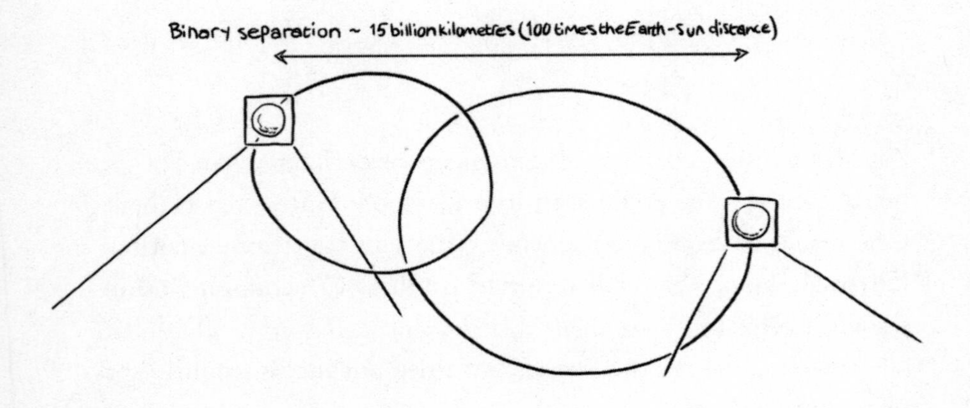

Twin star systems are not uncommon. It turns out that about 40 per cent of the stars in our galaxy are in binary systems – two stars circling each other.

All Sean Raymond had to do was position his two Ultimate Solar Systems away from each other by about 2.5 times the distance from Earth to Pluto – and they would be stable.

Voilá! Sixty potentially habitable planets in a single solar system.

"MINOR" PROBLEMS

Just to keep our simulations simple, I glossed over a few problems.

For example, I skipped over the possible destruction of the atmosphere of these Earth-like planets. This might happen as a result of them being so close to the powerful Solar Wind of a Red Dwarf.

Nor did I discuss the problems that can happen when you have similar-sized largish bodies (such as a pair of Earth-like planets) dancing around each other. These problems include massive tides, overwhelming tectonic plate activity, huge volcanism, rather long "days" of a few weeks or so (the much slower spin is due to gravitational locking), and more issues like these.

I also neglected another problem: How on Earth would we ever build such a Solar System?

The answer is quite straightforward. We'd need energy – lots and lots of it. In fact, more energy than we have ever had.

Back in prehistoric times, we could use only biological or animal energy. So we used the sweat of our brow, or that of a large animal.

The first real energy breakthrough came when we started using fossil fuels. Fossil fuels are incredibly concentrated energy. A barrel of oil contains about 159 litres of liquid. Translated to human terms, a barrel of oil holds the potential labour of a human working a 40-hour week for about four to ten years. At around US$100 a barrel, that's a bargain.

But if you are going to shift planets, that's not enough.

KARDASHEV CIVILISATION TYPE I

So let's look at the Kardashev Scale. It was first proposed in 1964 by the Soviet astronomer Nikolai Kardashev.

He defined a Civilisation not by its art or technological achievements, but by how much energy it could produce or use.

A Type I Civilisation would use as much power as its local star would send (in a thin pencil beam) towards its home planet. At our current rate of increase of power consumption, we will become a Type I Civilisation in a century or two.

The Sun dumps on our planet about 10^{16} Watts – that's 10 million billion Watts. If we want to generate that kind of power, we would have to do huge amounts of nuclear fusion. We would have to burn about 280 kilograms of hydrogen (turning it into helium and energy) each second.

That's a lot of power – but it's not enough to assemble planets from smaller asteroids, or move them around at our convenience.

KARDASHEV CIVILISATION TYPE II

Would we have to advance to what Kardashev defined as a Type II Civilisation to get the power we need to move planets?

To become a Type II Civilisation, we would have to be able to generate and control as much power as our Sun radiates in all directions. That's around 10^{26} Watts – or 100 trillion trillion Watts.

To generate that kind of power, our Sun burns about 620 million tonnes of hydrogen each second. At our current rate of power consumption increase, we will probably get to be a Type II Civilisation in a few thousand years.

But even that would not be enough.

KARDASHEV CIVILISATION TYPE III

We would need to become a Type III Civilisation.

According to Kardashev, that would mean being able to generate and control the amount of power put out by our entire Milky Way galaxy, with its hundreds of billions of stars. If you want a number, that's around 10^{37} Watts, or 10 trillion trillion trillion Watts.

At today's rate of power consumption increase, that would take us another few thousand years. That's about as far forward into the future as the early Egyptian civilisation is back in our past. On the other hand, if you want to account for the difficulties in handling such colossal amounts of power, it might be worth considering the more conservative approach of the theoretical physicist, Professor Michio Kaku and allow 100,000 to a million years to achieve Type III status.

Now I'm quietly confident that we will get through our current crisis with Global Warming – although it will be unnecessarily expensive, because we waited so long to do anything about it. But once we get past this Climate Crisis, I believe that one day we humans will control enormous energies. The jump we made from controlling the power of a horse to controlling the power of fossil fuels will seem puny.

One Cubic Metre of Higgs Field?

In July 2012, we proved that the strange property we call "mass" is caused by particles interacting with the Higgs Field.

This Higgs Field flows through everything – the vacuum of deep space, your body, and the centre of our Sun. The amount of Higgs Energy in just one cubic metre of "anywhere" is equal to the total power output of our Sun for 1000 years.

I suspect that our progress upwards along the Kardashev Scale will not be smooth and linear, but will go in sudden jumps.

For example, in 1905 we had no idea that Relativity would give us GPS or Sat-Nav. In 1972, we had no idea that the search for Black Holes would give us WiFi. In 100 or 1000 years from now, what will our descendants do with Dark Matter or Dark Energy? In 5000 years, will we be able to dump a star into a Black Hole and capture the power from the photons emitted from its accretion disc?

We know that enormous energies are there, but today, we have absolutely no idea how to access them. But what about our grandchildren, or their grandchildren, and so on?

Maybe they will be able to use starlight and really let it shine on their world . . .

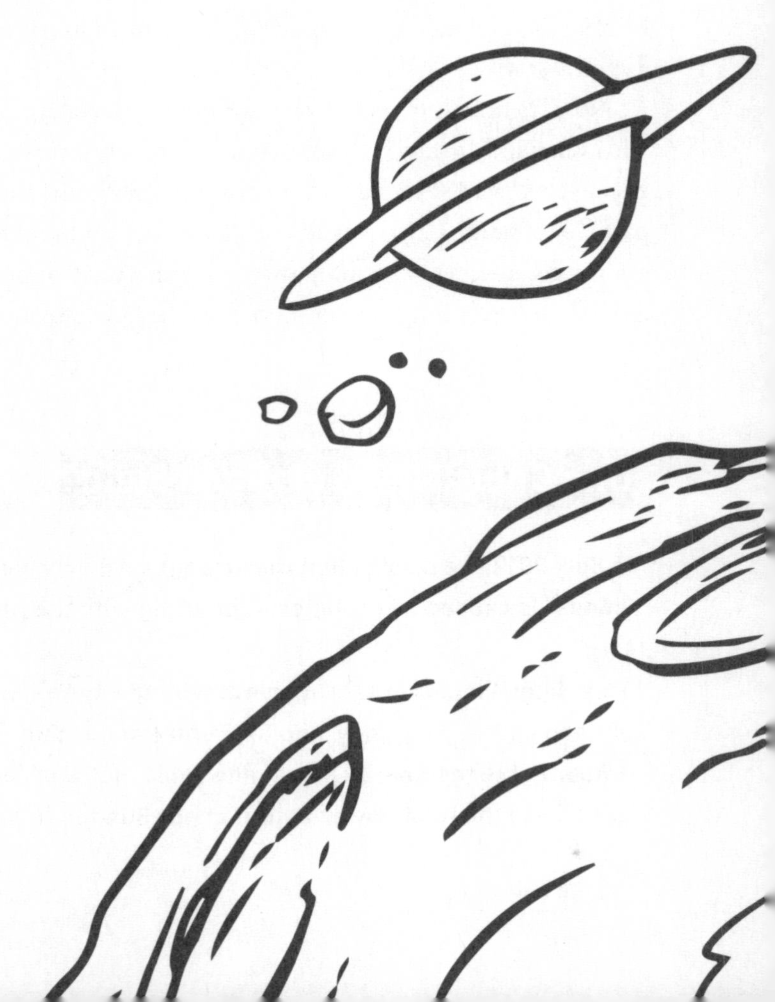

FOR
SALE

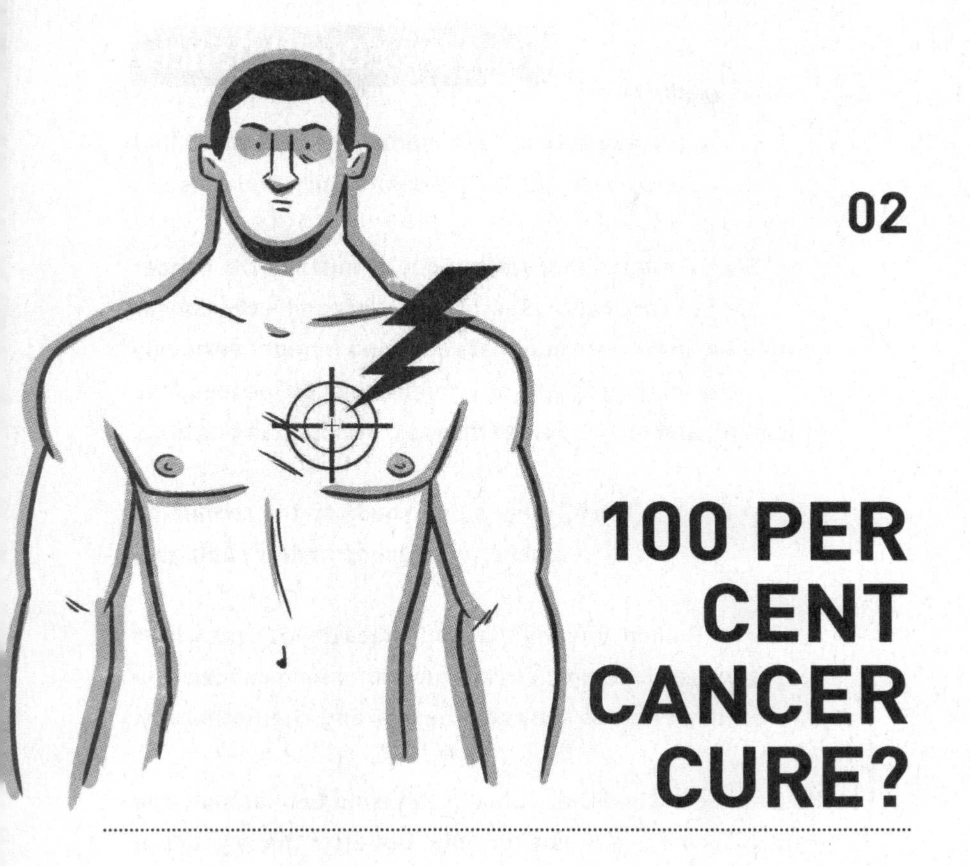

02

100 PER CENT CANCER CURE?

If there's one word you don't want to hear when you visit your GP, it's "cancer". Unfortunately, we don't seem to have a cure for all the various cancers. If you got a choice, you would pick a bacterial infection over cancer, because antibiotics can cure bacterial infections.

But, there is a very rare response to cancer treatments that somehow triggers a complete and permanent 100 per cent cure. This happens as a totally unpredictable – and let me emphasise, very rare – unexpected side effect of the treatment.

Welcome to the "Abscopal Effect". And yes, it's poorly understood – which is Scientific Talk for "we don't know the whole story, yet".

Treat (Not Cure) Cancer

In the majority of cases today, the treatment (not cure) of a cancer is "Slash and Burn and Poison".

"Slash" means that you cut out as much of the cancer as is practicable. Sometimes you can't – the cancer might be so widespread that removing it might seriously harm the patient, or it might be in an inaccessible location, and so on. Sometimes, a transplant is possible.

"Burn" means that you burn the remaining cancer with drugs and/or radiation.

"Poison" means that you inject the patient with a nasty chemical, again to kill any remaining cancer. The technical names are radiotherapy and chemotherapy.

With a cancer called CML (Chronic Myeloid Leukaemia), the survival rate used to be terrible. But after many years of painstaking research, we have fine-tuned the treatments. Today, nearly 90 per cent of treated patients will survive for five years. (This is the so-called Five-Year Survival Rate.)

We have recently begun Biological Therapy. We can now make large quantities of biochemical molecules that can sometimes destroy the new blood vessels that a growing cancer needs. Another line of research is the so-called "Tumour Vaccine", which can stimulate the patient's Immune System to find, attack and kill cancer cells.

IMMUNE SYSTEM VERSUS MICROBES

A while ago I came down with laryngitis. I wasn't worried in the slightest – although I did lie about with The Vapours for days, moaning quietly. I knew what would happen. My Immune System would recognise the bad guys that had invaded my larynx – and then kill them. Over a period of about a week I got worse, and then I got completely better, just as I expected.

My Immune System recognised the bad-guy microbes – and then killed them.

So why can't your Immune System recognise a cancer as the bad guy – and kill it?

The answer is that the cancer seems to be able to hide from and/or trick your Immune System.

But every now and then (so rarely that it falls into the category of the Genuine Medical Curiosity), the Immune System can recognise and kill an established widespread cancer.

This very rare phenomenon is called the Abscopal Effect. "Ab" means "away", while "scopal" refers to a "target". In medicine, the Abscopal Effect means that you treat a cancer in one part of the body and, somehow, all of that cancer anywhere else in the whole body is killed. Usually the treatment is a massive dose of X-rays.

CancerSSS, Not Cancer

There are many different Infectious Diseases. They range from the common cold to Ebola, and from the Flu to AIDS.

In the same way, there are many different cancers. There is no single disease called "cancer". (That's why I wrote "CancerSSS" – to emphasise that there are many of them.)

Rather, there are several hundred different types of cancers. They each have their own different cause, incidence and prevalence, pathology you see with a microscope, gross pathology that you can see with the naked eye, signs, symptoms, means of investigation, Natural History, treatment, and so on.

For example, consider the Natural History of two cancers.

Some 80 per cent of men will have cancer of the prostate by the age of 80. But in most cases it won't kill them, and they'll die with that cancer, not because of it. So cancer of the prostate can have a slow-moving Natural History. On the other hand, cancer of the pancreas moves much more quickly. It will kill 95 per cent of sufferers within five years.

Of course, there is another layer of complexity. For example, there are at least four different types of lung cancer, each with their own cause, cell type, aggressiveness, etc.

Cancer is a class of diseases. It is highly unlikely that there will turn out to be a single treatment for all cancers. (After all, there is not a single treatment for all Infectious Diseases.)

THE ABSCOPAL EFFECT, PART 1: ZAP CANCER

I first read about this in *Discover* magazine – their March 2014 issue. The story was called "Disappearing Act".

One day, Daniel, a healthy, fit man in his mid-30s, heard his basketball partner say the scary words, "Hey dude, what's that black blotch on the back of your shoulder?" Within a few weeks, Daniel was diagnosed with Malignant Melanoma.

This is a particularly aggressive type of skin cancer. A typical Five-Year Survival Rate is less than 10 per cent. This survival depends on how advanced the cancer is at diagnosis.

> Every now and then (so rarely that it falls into the category of the Genuine Medical Curiosity), the Immune System can recognise and kill an established widespread cancer.

A surgeon removed Daniel's single black mole. Unfortunately, it was too late. The Malignant Melanoma had already spread from its original location – before the surgery. Within two months, secondary Malignant Melanomas appeared in Daniel's lymph nodes. Within a year, more secondaries had sprouted in his lungs and liver.

At one year after diagnosis, he couldn't play basketball any more. He had lost about 15 kilograms, needed a cane just to walk and had also begun to suffer severe pain in his right hip. The hip pain was why he went to see Dr James Welsh, a radiation oncologist at the Midwest Institute for Neutron Therapy at Fermilab in Batavia, Illinois.

Unfortunately, the hip pain meant that secondary Malignant Melanomas had already spread into his bones, and had begun to cause major pain in at least one of them. Daniel's outlook was grim – he probably had only a few months of life left.

Dr Welsh decided to treat the hip pain with a high dose of radiation. Technically, this is called "palliative radiotherapy". The goal was purely to ease Daniel's hip pain – either by slowing down the growth of the secondary cancer in the hip, or even killing some of it. Dr Welsh couldn't give such a high dose of radiation to all the many secondary cancers throughout Daniel's body because such a treatment would kill him.

He made an appointment with Daniel for the week after the radiotherapy – but emphasised that if Daniel didn't feel well, he could simply ring and talk on the phone.

To Dr Welsh's surprise, Daniel returned in person one week later. The hip felt much better. This was "good" news – for the short term.

To quote Dr Welsh, "I scheduled him for a three-month follow-up, with the understanding that if he was not well enough to travel to the clinic, he needn't come in. Although I was thrilled his pain was gone, his prognosis was negative, and the palliative radiotherapy was not going to change that. Honestly, I did not expect to see Daniel ever again."

Pain and Death

I learnt many things in my time as a medical doctor.

The altruistic visionary Professor Fred Hollows taught me this gem: "Everybody has to die. What matters is that you have a good death."

A death free of pain is much better than a death with pain.

THE ABSCOPAL EFFECT, PART 2: CURE CANCER

Three months later, Daniel returned to Dr Welsh's practice. He looked much better. He weighed 5 kilograms more, and was so free of pain that he had stopped taking his pain medications.

Daniel insisted on having a CT scan. Amazingly, the many, many secondary cancers throughout his body (in his lymph nodes, bones, lungs, liver and so on) had vanished. All that was visible on the CT scan were tiny "radiographic scars", where the cancers had once been.

Cancer is a class of diseases. It is highly unlikely that there will turn out to be a single treatment for all cancers.

Another follow-up appointment was scheduled.

When Daniel returned another three months later, he was a different person. Not only had he regained a few more kilograms of solid muscle, he could now play basketball again and was back at work.

Fourteen years later, Daniel's Malignant Melanoma had still not returned. (I hope Daniel stays lucky.)

Let me emphasise very strongly that the Abscopal Effect is very, very, very uncommon. Don't expect it to be clinically available or promoted as a treatment option any time in the near future. It happens uncommonly in leukaemias and very rarely in other cancers.

It does seem that the Abscopal Effect is intimately involved and intertwined with the Immune System – which is fiendishly complicated. At the moment, our understanding of the Immune System is still very shallow and primitive.

It turns out that the field of Radiation Oncology is moving in new directions. One of them is the use of high-dose radiotherapy as a substitute for surgery – specifically to eradicate isolated secondary cancers. And very, very occasionally you get lucky and trigger the Abscopal Effect.

Spontaneous Regression

Spontaneous Regression, as applied to a cancer, means that the cancer just vanishes from the body. It's well documented. We don't fully understand it.

The five cancers to which this very rare phenomenon most commonly happens are renal cell carcinoma, lymphoma and leukaemia, neuroblastoma, carcinoma of the breast, and Malignant Melanoma (which is what Daniel had). In renal cell carcinoma, Spontaneous Regression happens in about 0.3 per cent of cases.

Perhaps the Abscopal Effect works in a similar way to Spontaneous Regression – and perhaps not.

I've been writing about "cancer" for a while without explaining exactly what it is. So let me fix that omission.

CANCER 101

Let me start with the several hundred different types of cells in your body – liver, lung, spleen, muscle, and so on.

As part of their life cycle, they grow. They all follow the general rule that they grow to a certain point – and no further.

But when something goes wrong with this process, the cell no longer follows that rule. (Mostly, we don't know why this happens.) Instead, that particular cell will keep on growing. It will invade other nearby tissues (something it would normally never do) and perhaps spread to distant parts of the body. It has become a cancer. By the time it has divided about 30 times, it will have generated about one billion cells and reached a mass of about one gram. At this stage, it is just clinically detectable.

In the year 2007, cancer killed nearly 8 million people worldwide. So cancer accounted for about 13 per cent of all human deaths.

The causes of cancers are many and varied – but, surprisingly, only about 5 to 10 per cent are inherited and can be passed from parent to child. So to some degree, some cancers are preventable. Diet, obesity and lack of exercise caused about 30 to 35 per cent of deaths directly attributable to cancer, followed by tobacco killing 25 to 30 per cent, infections at 15 to 20 per cent, and ionising radiation up to about 10 per cent. The different percentages vary from country to country.

Earliest Recorded Human Cancer

The earliest known record of a cancer is written in the Egyptian Edwin Smith Papyrus. It dates back to 1600 BC (so it's about 3600 years old) and describes a cancer of the breast.

CANCER VERSUS IMMUNE SYSTEM

Cancers are not "invisible" to your Immune System. They have a complex relationship with each other.

The Immune System tries to "edit" its own response over time. Its goal is to first stop the cancer from growing and, then, if the cancer has grown, to build an orchestrated response to it.

Think of this response as having three phases, which each begin with the letter "E" – Elimination, Equilibrium and Escape.

ELIMINATION

In the Elimination Phase, your Immune System sees the cancer and starts attacking it. This success of this process depends enormously on various factors – both local and distant. These include the specific cells the cancer started in, in which part of the body it's located, what kind of cells and chemicals your Immune System manufactures in response, and so on.

> In the year 2007, cancer killed nearly 8 million people worldwide. So cancer accounted for about 13 per cent of all human deaths.

But the cancerous cells have an advantage – they are continually mutating.

From a genetic point of view, our cells are relatively stable.

But cancer cells have several types of genetic instability seemingly built in – so their genome has enormous plasticity. Cancer cells keep on changing. Even if our Immune System recognises them, they change. This makes them effectively a moving target for the Immune System.

In general, your Immune System can wipe out, or eliminate, a percentage of the cancerous cells – but not all of them.

EQUILIBRIUM

After the initial Elimination Phase, you reach the Equilibrium Phase. A balance is established between the onslaught of the cancer and the defence mounted by your Immune System.

If you're unusually lucky, your Immune System will be able to eventually eliminate all the cancer cells – so the balance will drop to zero. If you're less lucky, you'll end up in a maintenance state, where the best you can do is keep the cancer the same size.

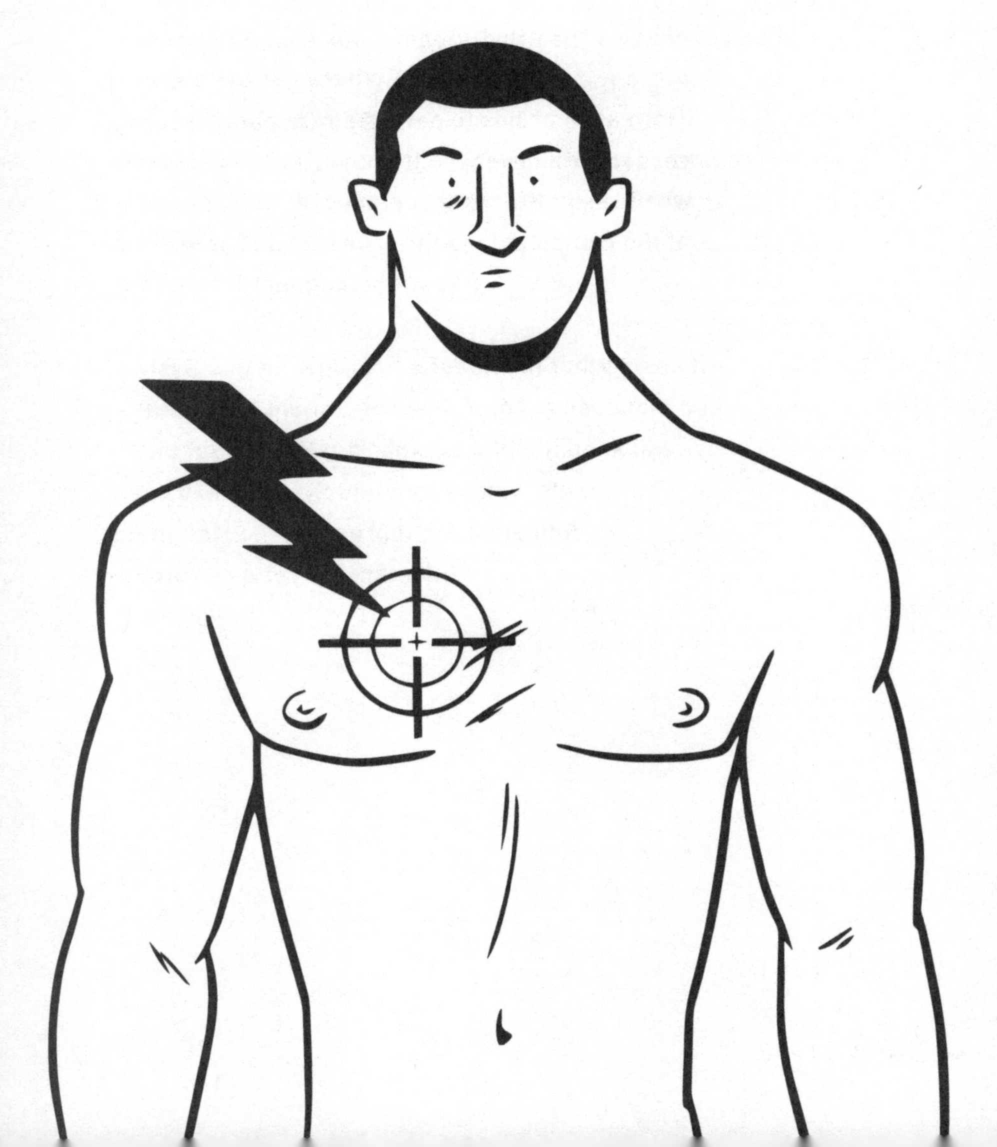

Escape from Equilibrium

A woman had Malignant Melanoma and was "successfully" treated for it. However, a small number of Malignant Melanoma cells survived unnoticed inside her kidneys – living in equilibrium with her Immune System. Sixteen years later, she died – for reasons completely unrelated to her Malignant Melanoma. She was evaluated to be "cancer-free" at the time of her death.

Each of her two "healthy" kidneys was transplanted into a willing recipient. Anybody getting a kidney from a donor has to have their Immune Systems suppressed with drugs, so that their Immune System won't reject the healthy kidney. Within two years of the transplant, each recipient had come down with Malignant Melanoma.

It seems that the donor's healthy Immune System had managed to control her Malignant Melanoma – and keep it in the Equilibrium Phase. But in the recipients, whose Immune Systems had been deliberately suppressed, the Malignant Melanoma cells ran amok.

ESCAPE

The Escape Phase is less exciting than it sounds, given that it's the cancer that escapes, not you!

Here, the Immune System can no longer regulate the cancer cells, which escape from its control – and start spreading.

THE ABSCOPAL EFFECT "EXPLAINED"

So what's happening on those incredibly rare occasions when the Abscopal Effect kicks in?

Here's a rough overview.

When the local cancer (say, in Daniel's right hip) gets zapped with a large dose of radiation, it protests. It dumps massive amounts of foreign chemicals into the bloodstream. The Immune System suddenly responds to the huge exposure to the "insides" of the dead cancer cells. Part of the response involves Killer T-Cells.

This is the pathway that, in Daniel's case, led to *all* of the Malignant Melanomas inside his body vanishing (or "regressing").

But there are so many fine details that we don't know. Why does it happen only rarely? How can we make it happen 100 per cent of the time, for all the different cancers?

There is so much more to learn, and none of us should be immune to that . . .

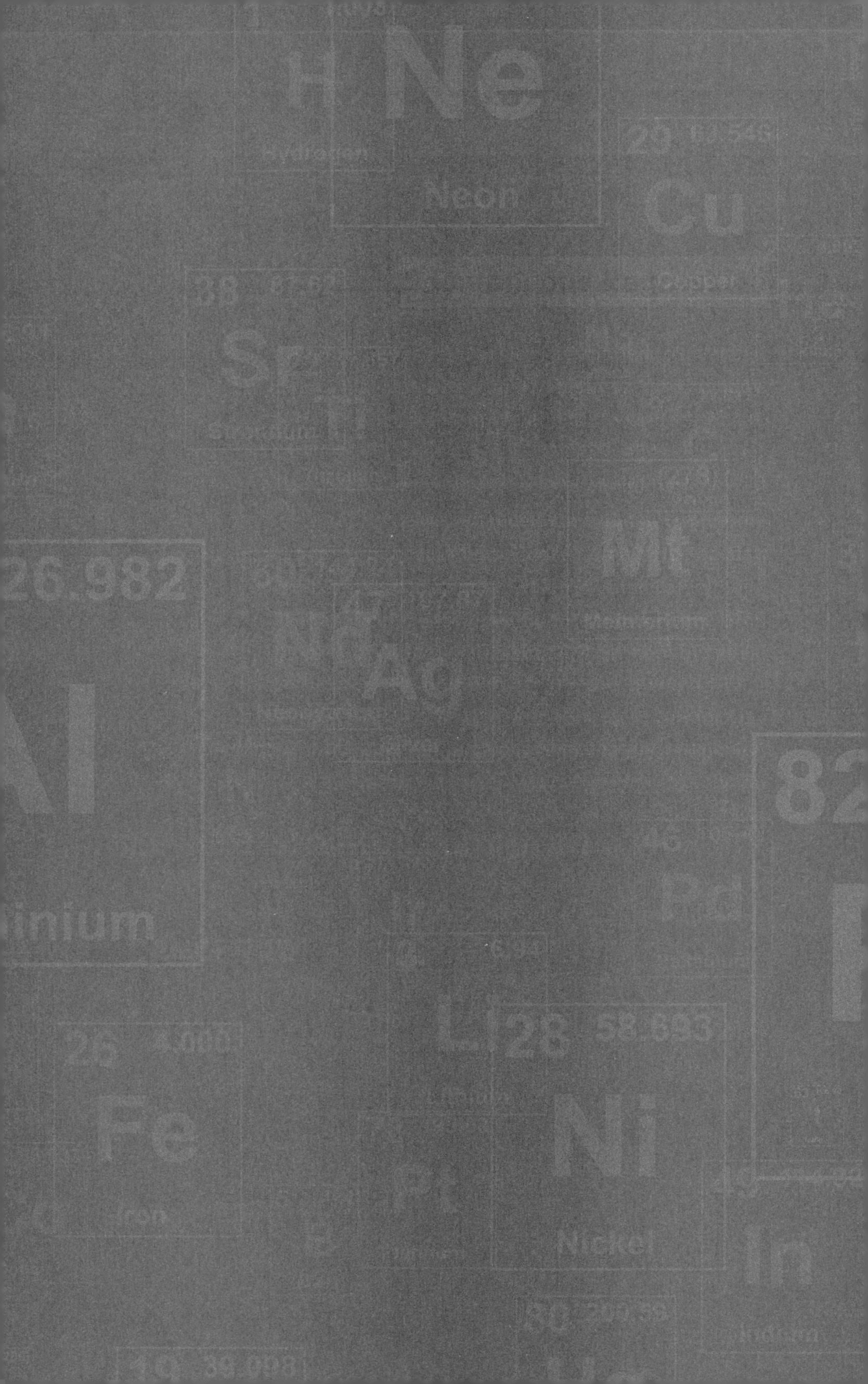

03

AHEAD OF HIS TIME

"Time" is something that both philosophers and physicists have been wondering about for, well, a very long time. And now, the neuroscientists have joined in the discussion.

They have found a man who hears what you are saying to him before he sees your lips move.

TIMING IS ALL

Usually, when you look at a ball bouncing on the ground, or the face of a person speaking to you, the "sight" and the "sound" are synchronised. You perceive the sight and sound as happening at the same time.

> In terms of vision you are always living three-tenths of a second (0.3 seconds) behind reality.

But not so for a man known only by the letters PH, who is a retired pilot, and a highly intelligent man. People usually see and hear the world in sync. Like the rest of us, PH used to do that, too.

But in November 2007 he had surgery for pericarditis – an infection of the covering of the heart. In 2008 he developed generalised myaesthenia gravis. This is an autoimmune disease that leaves you with muscle weakness that comes and goes – and you also get easily tired. (We don't know if either of these two events had anything to do with what came next.)

About two to three months after the onset of the myaesthenia gravis, things got weird. It happened rather suddenly.

Mr PH first noticed something strange about his daughter's two TV sets. He was hearing people on the TV speak before he saw the lips move and he said to his daughter, "Hey, you've got two TVs that need sorting!" But the TVs were fine.

He suddenly realised that the same mis-synchronisation was happening in conversation with people – first he heard the voice, later he saw the lips move. Then he noticed that he could hear his own voice before he felt his own jaw move.

Something odd had happened with the audio timing circuits inside his brain.

VIDEO PROCESSING = 0.3 SECONDS

You might not realise this, but in terms of vision you are always living three-tenths of a second (0.3 seconds) behind reality. That's the "processing time". Yup, vision is not instant.

The process starts when the incoming light lands on the cells in your retina. It finishes with the glorious full wraparound 3-D colour sensation we call vision.

In between, a lot happens inside your head.

First, the rods (at night) or the cones (in daylight) in your retina turn the incoming light into electricity. The remaining nine or so layers of cells in the retina then process and compress that electricity. At the end of this initial process, your retina "knows" the difference between a horizontal bar in your field of view and a vertical bar. Furthermore, your retina "knows" if it's a white bar on a black background or vice versa.

Then the compressed signal leaves the retina and gets sent along the optic nerves to the visual cortices in your brain right at the back of your skull. The visual cortices are called Brodmann Areas 17, 18 and 19.

The final stage is that the visual cortices decompress and re-process the electrical information to give you the magnificent experience of vision. This all takes a significant amount of time – about three-tenths of a second, or 0.3 seconds.

Normally, we don't notice this delay, because we have evolved to be able to deal with it. We compensate for this delay by anticipating actions that we have seen before.

AUDIO PROCESSING – 0.1 SECONDS

With hearing, as with vision, you are living a fraction of a second behind reality.

Just like our "video" signal, our "audio" signal has to be processed. It takes time for the incoming audio signals from our ears to be processed.

If you've ever dealt with audio and video computer files, you would know that the audio files are much much smaller than the video files. The same seems to also hold for human audio and video files.

In the case of the human brain, when the Brodmann Areas associated with hearing (Areas 41 and 42) start processing the incoming audio information, they do so quite quickly, because the amount of data is rather small.

Remember, audio processing takes one tenth of a second, while video processing takes three tenths of a second.

AUDIO PROCESSING + DELIBERATE DELAY

So in conversation, when we stand close to someone, how come we see the lips moving at the same time as we hear the voice? Almost certainly it's because our brain deliberately inserts a delay of about two-tenths (0.2) of a second on the audio circuit.

You would immediately think this could be a disadvantage. If something dangerous, such as a killer rabbit, is charging you at full speed, you want to know about it as soon as possible. That extra two -tenths of a second could save your life.

But perhaps evolution "decided" it was more important for us to be able to easily communicate with our fellow humans than to hear that killer animal 0.2 seconds sooner.

After all, the individual human is quite pathetic as a hunting animal. We can't sprint very quickly, our vision is not very sharp, and neither is our sense of smell. (But we are good long-distance runners. See "Dancing With Deer Evolves Your Brain", page 85) Our claws are silly little fingernails, and our teeth are not very good at ripping and tearing. But thanks to our amazing brain, we can organise ourselves into groups and so we have become "Masters of the Planet". Perhaps being able to communicate easily with each other helped us in this process.

> With hearing, you are living just one tenth of a second behind reality.

The neuroscientists did an MRI scan on PH's brain, and found damage in areas that were "well-placed to disrupt audition and/or timing". That's why Mr PH has lost the 0.2 second delay that the rest of us experience so that we can enjoy sounds and sights in sync. The neuroscientists tested him by deliberately delaying sound by two-tenths of a second. Suddenly audio and vision were in sync for him again.

Today, he can avoid killer rabbits just that little bit sooner than the rest of us, but conversation is a little odd.

I wonder if Mr PH has an advantage over the rest of us with witty one-liners . . . or is that just a concept he pays lip service to?

Fix Mr PH's Problem

I can think of two "easy" ways to fix Mr PH's timing problem.

One is that he wears special headphones all the time. The headphones would delay the sound by two tenths of a second, before releasing it to land on his ears. (Okay, not very practical.)

Another is that he has conversations from 60 metres away. (Okay, again, not very practical.) Sound takes about two-tenths of a second to cover 60 metres. Unfortunately, he would need binoculars to see the other person's lips – and a microphone and loudspeaker.

04

APPLE LOGO AND ALAN TURING

Big corporations are becoming, for better or for worse, an increasingly powerful force in today's world. Most of us will recognise them by their logos – the curly Coca-Cola typeface, the stark red emblem for the Red Cross, or the simple fruit logo of the technology gargantuan known as Apple.

Rumours have circulated about why the original popular Apple logo that lasted over two decades was a rainbow-striped silhouette of an apple – and why the apple had a bite out of it. Is it because a persecuted gay genius invented the computer?

Logo 101

"Logo" is an abbreviation of "logotype", which in Greek literally means "word imprint".

So today a logo is some kind of mark, emblem or symbol by which you can instantly recognise commercial organisations. A logo can be a purely graphic symbol (such as the "swoosh" of Nike), a name (such as the letters "I", "B" and "M"), or a combination of symbols and letters.

ORIGIN OF THE APPLE LOGO

The story most commonly told about the Apple logo comes from World War II, and does indeed involve a genius. His work, according to both British Prime Minister Winston Churchill and Supreme Allied Commander Dwight D. Eisenhower, enabled the Allies to win World War II.

But, first, why the bite out of the Apple logo?

Is it a play on words because, in computer terms, a "byte" is a unit of data? Is it related to the Apple of Knowledge in the Garden of Eden with the two naked people and the talking snake? Is it because Isaac Newton supposedly got hit on the head by a falling apple, and so devised his Theory of Gravity? Or is the bite there because in various societies the apple has been linked to concepts of lust, knowledge, anarchy and hope?

No, none of these.

Well then, what about the story that the rainbow Apple logo had a bite out of it because of Alan Turing?

Apple's Very First Logo

Apple's very first logo was used for only one year – 1975.

It was a rectangular-framed pen-and-ink line drawing of Newton sitting under an apple tree, with the words "Newton – A Mind Forever Voyaging Through Strange Seas of Thought – Alone" running around the border. That famous apple had just begun to fall downwards. It was too complex to become iconic, not to mention a bit of a downer.

It was designed by Ronald Wayne, who was, very briefly, a partner in Apple. He was the only partner who had a family, and felt he needed more financial security. Just a few weeks after helping found Apple, he sold his shares for US$2300. Today, those shares are worth about US$35 billion (depending on the current market capitalisation of Apple).

ALAN TURING – GENUINE GENIUS

Turing was born in 1912 and was in every sense a true genius.

> Alan Turing invented the concept of a "Universal Machine". This machine would be able to do anything.

By coincidence, his first day of high school coincided with the 1926 General Strike in Britain. As part of the General Strike, public transport networks were closed down. Undeterred, he rode his bicycle by himself about 100 kilometres, stopping overnight at an inn.

By the age of 16 he had read and understood Albert Einstein's writings on relativity. He had even taken Einstein's work further in the context of questioning Isaac Newton's Laws of Motion.

Specifically relevant to our story, in 1936 Alan Turing also invented the concept of a "Universal Machine". This machine would be able to do anything. According to Turing, this theoretical machine should be able to get a number (say, off a paper tape) and then either add it to or subtract it from another number (somewhere else on that same tape), and then invert or turn upside-down any of these numbers or results and, lastly, store the final result – again on this tape.

If it could do that relatively "simple" operation, it would be Turing's beloved Universal Machine.

Turing the Eccentric?

Turing was prone to hay fever. When pollen season came around he would continue cycling to work, but wore a military gas mask to keep the pollen out.

His bicycle had a problem – the chain would come off. Turing soon realised that it came off after a fixed number of revolutions. Did he fix the chain-and-cog assembly? No. He simply counted the revolutions of the pedals. When he got to just before the number when the chain would come off, he would stop, manually adjust the chain and continue cycling.

He also didn't like his tea mug in the tea room being stolen or used by others. So he chained it to the radiator pipes.

UNIVERSAL MACHINE = COMPUTER

I first really understood the concept of Alan Turing's Universal Machine when I walked 790 kilometres across Spain on the El Camino de Santiago de Compostela in 2009. It took about a month. I realised that my Smartphone, which easily fitted in my pocket, was what Alan Turing dreamed of in his Universal Machine.

Way back in 2009, the Smartphone was still a new concept.

My Smartphone, of course, worked as a phone.

But wait, there's more . . .

It was also a camera, and a GPS device that showed me The Way – as well as a talking Spanish–English dictionary. Being Smart, it gave me access to email and the web. And it was also my diary and notebook, calendar and calculator, currency converter and clock, and a map of the night sky and a compass. It also gave me access to daily weather forecasts, newspapers anywhere in the world, and was a voice recorder.

Truly, my Smartphone was a computer that, just like Turing's Universal Machine, could do anything.

Data = Applications

There was another surprising realisation that came from Turing's work.

In the real world, there is a big difference between a tool and the stuff that the tool acts upon. For example, in carpentry, a metal saw (a tool) is quite different from the timber (which is processed by the saw).

Not so in the virtual world, and Turing realised this.

He proved that there is no fundamental difference between "data" and the "instructions" that process the data. (Today, we label the "instructions" as "programs" or "applications".)

Today, we all accept that our computers easily handle both programs and data as bunches of 1s and 0s. But Turing was the first person to *prove* mathematically that there was no essential difference between data and applications. This deep and fundamental understanding sped up the introduction of computers by decades.

SLOW BIRTH OF THE COMPUTER

By 1941, the number of possible encryption codes of the German Navy's Four-Rotor Enigma Machines was about 180 billion billion.

The modern computer had its origin before World War II, when Turing was already working part time with the British Government Code and Cipher School. World War II broke out in 1939 and, within a frighteningly short time, the Nazis had conquered mainland Western Europe.

The UK was (to some degree) protected from invasion by the English Channel. When the USA entered the war, it saw the UK as an "unsinkable aircraft carrier". In other words, to the North Americans it was a potential offshore base from which mainland Europe could be one day liberated.

The USA was sending supplies across the Atlantic to the UK via high-speed sea convoys. Surprisingly, the much slower Nazi submarines (called U-boats) repeatedly and consistently sank the fast convoys – up to 400,000 tonnes of shipping each month.

How?

Because the Nazi High Command guided the *slow* U-boats to locations where the *fast* convoys could be intercepted – via coded radio messages. These messages were encoded by a family of German encryption machines collectively known as "Enigma".

Turing helped develop a machine (the "Bombe") specifically to bust Enigma encryption. It was the predecessor of the modern computer. The British government concentrated all their major de-encryption work at Bletchley Park, about 64 kilometres from London.

Computer Was a Person

Today, we think of a "computer" as a machine.

But before Turing, a "computer" was a person. They were usually women, and were hired to do repetitive calculations for hours and days on end.

The first "computers" were Alexis Claude Clairaut and his two fellow astronomers. Back in the 1750s, Clairaut was trying to plot the path of Halley's Comet. He and his colleagues worked out where the comet was and how fast it was moving. They then used Newton's Laws of Motion to work out its location and speed at a time (say) one minute later. They then repeated the whole procedure again, and again, and – you get the idea.

It took them five months to work out the path of Halley's Comet as it zipped past Saturn and Jupiter. Their estimate of the point of closest approach of the comet to the Sun was quite accurate.

ENIGMA 101

The German Enigma machines had a truly sophisticated design. Enigma wasn't a single machine – it was a whole family of slightly different machines that evolved over a few decades.

Enigma was originally invented at the end of World War I by the German engineer Arthur Scherbius. The major mechanical parts were a keyboard, a set of rotating discs called "Rotors" and a Plugboard. Both the Rotors and the Plugboard could swap one letter of the alphabet for another.

The Enigma encryption was based on the concept of a Substitution Code.

Suppose you had a Substitution Code which just swapped letters from one end of the alphabet to the other – A = Z, B = Y, and so on up to Z = A. That would be fairly easy to break. Of course, you could make the Substitution Code more complicated than this. Even so, somebody would eventually be able to break it, if they could get enough encoded messages to examine – and find a pattern.

You might, for example, use the knowledge that the most commonly used letters in the English language are "e", "s" and "t". So there's a good chance that the most commonly appearing characters in the encoded message stand for "e", "s" and "t".

But suppose that you change the Substitution Code with each line of text you send. That would make the message harder to decode. It would be even harder if you changed the Substitution Code after each word was sent.

ENIGMA – RESISTANT TO "BRUTE FORCE"

The brilliance of Enigma was that it changed the Substitution Code after each character was encoded. If you pressed "a" three times in a row, Enigma would produce three different letters, such as "g", "z" and "b".

At the beginning of each day, the sender and receiver would set the Rotors and Plugboard to the same settings. They would get the settings from a Settings Book that was sent out each month. For the rest of that day, the two Enigma machines would spin their Rotors to change their Substitution Codes in lockstep with each other.

There were many versions of Enigma. The more Rotors they used, the harder they were to bust. By 1941, the number of possible encryption codes of the German Navy's Four-Rotor Enigma Machines was about 180 billion billion. If all you had were the encoded messages, there was no way you could de-encrypt them with a "brute force" attack – just by trying every single possible combination.

Healthy Mind in a Healthy Body

Alan Turing was also a talented long distance runner. Bletchley Park was some 64 kilometres from London – and Turing would sometimes run the distance there for meetings.

After World War II, he tried out for the 1948 British Olympic team, even though he was slowed by an injury. In the 1948 Olympics, in the marathon, his time was only 11 minutes slower than that of fellow British team member Thomas Richard's silver medal win, which was 2 hours and 35 minutes.

ALLIES BREAK ENIGMA

How did the British de-encryption teams, based at Bletchley Park, break Enigma? On one level, they designed and built special de-encryption machines, called Bombes (more on them later). But there were many factors, mostly involving plain dumb luck, which helped the British.

First, the Polish Cipher Bureau did a lot of the groundwork. They had actually broken an early version of the Enigma code in December 1932. This was thanks to brilliant mathematicians, and it didn't hurt that they had a spy inside the German office who took care of encrypted communications.

In 1938, the Poles had designed and built a machine that could de-encrypt the earlier and more simple Enigma codes. They called it a "Crypotologic Bomb". But then the Germans made the Enigma machines 10 times more complex. The Poles could not supply 10 times more machines and personnel for this task. In the lead-up to World War II, it was beyond their capabilities.

On 25 July 1939, just five weeks before the start of World War II and the German invasion of Poland, the Polish Cipher Bureau could see what was coming. They handed over to the Allies a virtual goldmine – reconstructed Enigma machines, and the mathematical techniques for de-encrypting the coded messages.

Second, Enigma had a fundamental flaw. It could never encode a letter of the alphabet as that same letter – "a" would never encode to "a". This gave the people at Bletchley Park a bit of leverage for busting the Enigma code.

Third, the German operators got sloppy. The German Air Force provided the most clues to Bletchley Park, because compared to the Army and Navy, their operators were especially undisciplined. At the beginning of each day, the standard procedure was to send a test message to each other to make sure their machines were properly set up with the same Rotor and Plugboard positions. The test message was supposed to be different each day. The operators got lazy, and sometimes sent the same message each day – for example, that day's date followed by "*Heil* Hitler". Again, this gave Bletchley Park slight leverage.

Fourth, the operators occasionally made procedural flaws. In December 1941, the German Navy was soon to officially introduce the new Four-Rotor Machine. This would have made it virtually impossible for the teams at Bletchley Park to bust the code. Fortunately for the Allies, a U-boat radio operator made a mistake. He accidentally broadcast a message using the Four-Rotor Enigma, and then immediately broadcast the exact same message using the Three-Rotor Enigma. This act revealed to Bletchley Park the wiring of the fourth rotor – so they could (with extra work) decode it.

In 1946, Turing presented a paper that gave the first detailed design of a stored-program computer, similar to our computers today.

Fifth, the Allies were fortunate enough to get hold of a few current Enigma machines – and the current code settings for that month. In May 1941, the British captured a Naval Enigma machine from the German submarine *U-110*. They also captured *U-559*, seizing current Enigma codebooks and settings, on 30 October 1942. On 4 June 1944, American forces captured another Naval Enigma machine, as well as current codebooks and Enigma settings from *U-505*. These cryptographic materials arrived at Bletchley Park on 20 June 1944.

Having the code settings meant that the British could immediately process encrypted communications with their ill-gotten Enigma machines. Bletchley Park didn't just have to use their overloaded Bombes.

Secrets Behind Secrets

If the Nazi High Command had realised that their supposedly secret Enigma communications were regularly being decoded, they would have made the Enigma machines more complicated. They could have done this quite easily by adding more Rotors, and more Plugboard connectors.

To stop this from happening, the Allies played dumb. Even though the Allies knew (thanks to the work of Turing and others at Bletchley Park) that certain German supply ships were travelling along a specific pathway to a specific location, they did not simply send out bombers to destroy them. To cover up their secret knowledge, the Allies sent out several reconnaissance aircraft – only one of which would "find" the German supply ships, and only then call in the bombers. The other reconnaissance aircraft were deliberately sent on futile missions, so that the Nazis would not suspect that their secure Enigma system had been broken.

THE BOMBE

In 1939, Turing and his team at Bletchley Park examined the re-constructed Enigma machine given to them by the Polish mathematicians.

They used the knowledge they gained from this analysis to design an improved (but still primitive) electromechanical de-encryption machine they called a Bombe. The first one was up and running at Bletchley Park on 18 March 1940. The Bletchley Park people were able to decode the daily traffic of the Nazi Enigma machines quickly enough for the information to be useful from a military point of view.

By the middle of 1941, the Bombes had evolved into machines about 2 metres high, 2 metres wide, 60 centimetres deep, and weighing a tonne. They could de-encrypt three Enigma codes at the same time. Bletchley Park was decoding 1000 Enigma codes each day.

There weren't enough Bombes, or people to use them. But even so, they were very successful. By July 1941, they managed to reduce the shipping losses of the high-speed convoys to 80,521 tonnes per month.

By the end of the war, there were about 200 of these electromechanical Bombes noisily clacking away, decoding supposedly secret Enigma messages.

The Bombe was not a true Universal Machine. It was not "programmable" to do many different jobs, like our modern computers. It could do only one task. But it was an essential evolutionary step along the way.

The work of Turing, and his colleagues at Bletchley Park, was kept secret for over three decades.

Secrets After World War II

Why did the British Government keep Turing's work secret after World War II? There are three main theories. Feel free to pick one, or simply make up your own.

First, the British Government was selling encryption equipment to foreign governments around the world. Indeed, they had grabbed all the Nazi Enigma machines they could find (over 100,000 had been built). The British were selling them to poor countries, which thought that they were still secure. So it was in the British Government's

interest to keep it secret that they had the ability to bust the encrypted messages of foreign governments.

Second, some years earlier, Winston Churchill had apparently publicly revealed information from de-encrypted Soviet messages. The Soviets realised that their encryption techniques were broken, and promptly changed them. Suddenly, the British could no longer read Soviet communications – for a while. This theory claims that Turing's work at Bletchley Park was kept secret, so that the British could continue to snoop on the Soviets.

Third, on 25 May 1945, apparently Winston Churchill specifically asked all those who benefitted from the work at Bletchley Park to keep it secret. This would allow the British Secret Service to continue snooping.

Regardless of which theory is correct, the French Intelligence Officer Gustave Bertrand published the story of Enigma in his book *Enigma* in 1973. The British finally lifted the ban on talking about Bletchley Park's work in 1974 – too late for Turing.

TURING AFTER WORLD WAR II

So, unlike the people who worked on the Atom Bomb in the USA and were made famous, all the British cryptographers were completely unknown. In fact, they were still bound by the *Official Secrets Act*.

After World War II, Turing and his colleagues worked on the first true Universal Machines – machines that today we would call computers. In 1946, he presented a paper that gave the first detailed design of a stored-program computer – similar to our computers today. (Think how many stored programs are on your computer. They include web browsers, word processors, Skype, Photoshop, address books, spreadsheets, GPS and maps, CD and DVD burner applications, scanner programs, presentation programs, music players, video and movie players, etc, etc.) In 1948, he worked at Manchester University in developing actual computers.

Turing then shifted his intellectual focus – he began trying to decode Life itself. He started working on Mathematical Biology – for example, how did the fingers on your hand get formed or how did the zebra get its stripes? Remember, this was before DNA had even been discovered. He was so far ahead of his time that his work has only recently been proven to be on the right pathway.

TURING COMMITS SUICIDE

In January 1952, Alan Turing's house was robbed. He told the police that he suspected the robber to be a friend of his boyfriend.

At that time, homosexuality was a criminal offence in the UK. Alan Turing was convicted and given the choice of either imprisonment or probation with chemical castration – that is, injections of female hormones. Alan Turing chose the latter punishment of the feminising hormones, became sexually impotent, and grew breasts. He was labelled a Security Risk, and no longer had access to the people and projects he loved.

On 7 June 1954, he died, by apparently committing suicide by eating an apple laced with cyanide. Indeed, a half-eaten apple was found in his apartment.

What is the significance of the half-eaten apple?

It turns out that Turing's favourite movie was *Snow White*. You know the plot – the Princess falls into a deep sleep after eating the poisoned apple, and is awakened by the kiss of the Prince.

So that's the theory why Turing committed suicide with a poisoned apple.

RAINBOW BITE APPLE LOGO

Some two decades further on, is that the reason why the Apple company designed their logo to have a missing bite – to honour Alan Turing?

And was it as a salute to Gay Pride that the Apple logo from 1976 to 1999 also incorporated a horizontal band of rainbow colours?

NO, SORRY, THAT'S WRONG

The story of Apple honouring Alan Turing is very moving and tragic, and widely repeated. However, it's an Urban Myth, and incorrect in a few details. But, like all successful myths, it has elements of truth in it.

Stephen Fry asked Steve Jobs, one of the founders of Apple, whether the Urban Myth was true. Steve Jobs replied, "God, we wish it were."

Yes, at the inquest the coroner determined that Alan Turing did die from cyanide poisoning. Yes, there was an apple with a bite taken out of it in his apartment.

But first, the coroner did not analyse the apple to see if it contained cyanide. So we'll never know that detail.

Second, there is circumstantial evidence suggesting it was not a suicide, but an accident. Alan Turing had set up an apparatus to electroplate spoons with gold in his tiny spare room. He was using potassium cyanide to dissolve the gold and may well have accidentally inhaled cyanide fumes. He always ate an apple before going to bed, and would sometimes leave them half-eaten.

Despite the stress of his legal setbacks, and the side effects of the female hormones, Turing was not depressed and indeed was in good humour. In fact, he had written down a list of jobs to complete at work after the holiday weekend.

LOGO DESIGNER TELLS ALL

In 1999, *Time* magazine designated Alan Turing as one of the 100 Most Important People of the 20th Century, and wrote, "The fact remains that everyone who taps at a keyboard, opening a spreadsheet or a word-processing program, is working on an incarnation of a Turing Machine."

On 10 September 2009, the British Prime Minister Gordon Brown, on behalf of the British Government, made an official public apology for "the appalling way [Turing] was treated". On 24 December 2013, the Queen granted him a posthumous pardon.

In 2009, Rob Janoff, who actually designed the famous Apple logo, explained how he did it. The only direction that he got from Steve Jobs at the time was, "Don't make it cute."

So Rob Janoff tried many concepts. But he kept on coming back to a straight representation of an apple. The trouble was that in a simple line drawing, an apple looks like a cherry. So in Rob's exact words, "I designed it with a bite for scale, so people get that it was an apple not a cherry. . . . It was after I designed it that my creative director told me: 'Well you know, there is a computer term called byte.' And I was like: 'You're kidding!'"

And the rainbow stripes? Again in Rob's exact words, "the real solid reason for the stripes was that the Apple II was the first home or personal computer that could reproduce images on the monitor in colour. So it represents colour bars on the screen."

So think of this story every time you tap your Smartphone's screen "Turing" somebody . . .

05

ARCTIC MELTDOWN – MILANKOVITCH CYCLES

In the Arctic, the volume of the ice changes with the seasons. It rises to a peak after midwinter, and falls to its minimum after midsummer – usually in the month of September.

Over most of the past 1400 years, the volume of the Arctic ice each September has stayed pretty constant. But something has changed recently. Since 1980, we have lost 80 per cent of that summer ice.

That's correct – not 8 per cent, or 18 per cent, but 80 per cent!

MILANKOVITCH CYCLES

> The Milankovitch Cycles control how much solar radiation lands on the Earth, secondly, whether it lands on ice, land or water, and finally, when it lands.

Over the last 4.7 billion years, there have been many *natural* cycles in the climate – both heating and cooling.

What's happening today in the Arctic is not a cycle caused by nature. This more recent change is caused by us humans. We've burnt fossil fuels and dumped slightly over one trillion tonnes of carbon into the atmosphere over the past century.

But what are these natural cycles? There are many cycles that influence our climate – short-term, medium-term and long-term. But let's look at just one of the medium-term cycles – the Milankovitch Cycles. They're named after the 20th century Serbian scientist, Milutin Milankovitch.

These cycles relate to the Earth and its orbit around the Sun. There are three major Milankovitch Cycles. They interact with each other. Between them, they have three major effects on the energy that comes from the Sun. They first control how much solar radiation lands on the Earth, secondly, whether it lands on ice, land or water, and finally, when it lands.

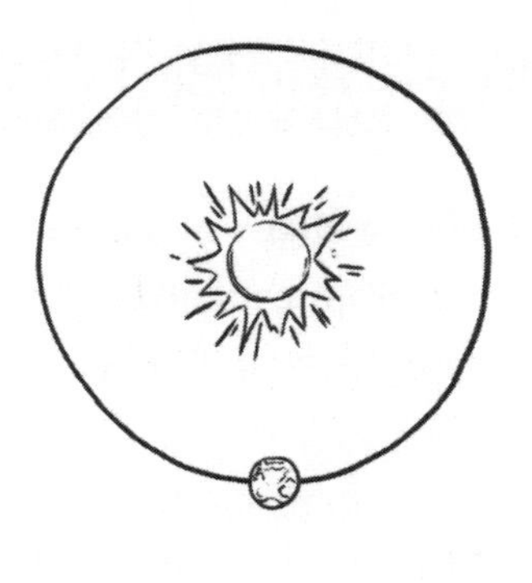

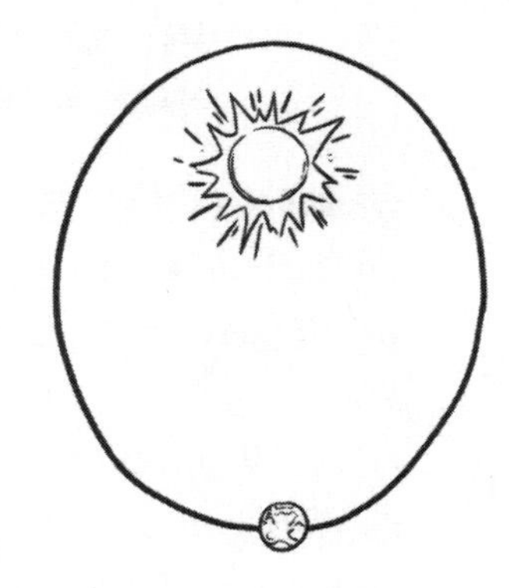

MILANKOVITCH CYCLE 1: CIRCULAR TO ELLIPTICAL

The first Milankovitch Cycle is that the orbit of the Earth changes from "nearly circular" to "slightly elliptical". It does this on a predominantly 100,000-year cycle.

When the Earth is close to the Sun it receives more heat energy, and when it is further away it gets less. At the moment, the orbit of the Earth is about halfway between "nearly circular" and "slightly elliptical". So over each calendar year, the change in the distance from the Earth to the Sun changes by about 5.1 million kilometres. This translates to about a 6.8 per cent difference in incoming solar radiation.

But in 50,000 years, the orbit of the Earth will be at its most elliptical. The change in distance between the Sun and Earth will be much greater during each calendar year. As a result, there will be a 23 per cent difference over the year in how much solar radiation lands on the Earth.

Elliptical to Circular, Repeat

When I said that the cycle for the shape of the Earth's orbit to change from circular to elliptical and back took about 100,000 years, I was giving the simplified version.

As you would expect of anything in our Universe, it's more complicated than that. It turns out that there are other elliptical–circular cycles that have periods of 413,000 years, 95,000 years and 125,000 years. But when you account for their various "strengths", they loosely combine to give a predominantly 100,000-year cycle.

MILANKOVITCH CYCLE 2: THE TILT FROM VERTICAL

The second Milankovitch Cycle affecting the solar radiation landing on our planet is the tilt of the North–South spin axis. This is usually compared to the plane of the orbit of the Earth around the sun.

This tilt rocks gently between 22.1° and 24.5° from the vertical. This cycle has a period of about 41,000 years.

At the moment we are roughly halfway in the middle – we're about 23.44° from the vertical and heading downwards. As we slowly head down to the minimum of 22.1° – reaching it around the year 11,800 – there will be three major trends.

First, the summers in each hemisphere will get less solar radiation. Second, the winters will get more solar radiation. Averaged out over the planet, there will be a slight overall cooling.

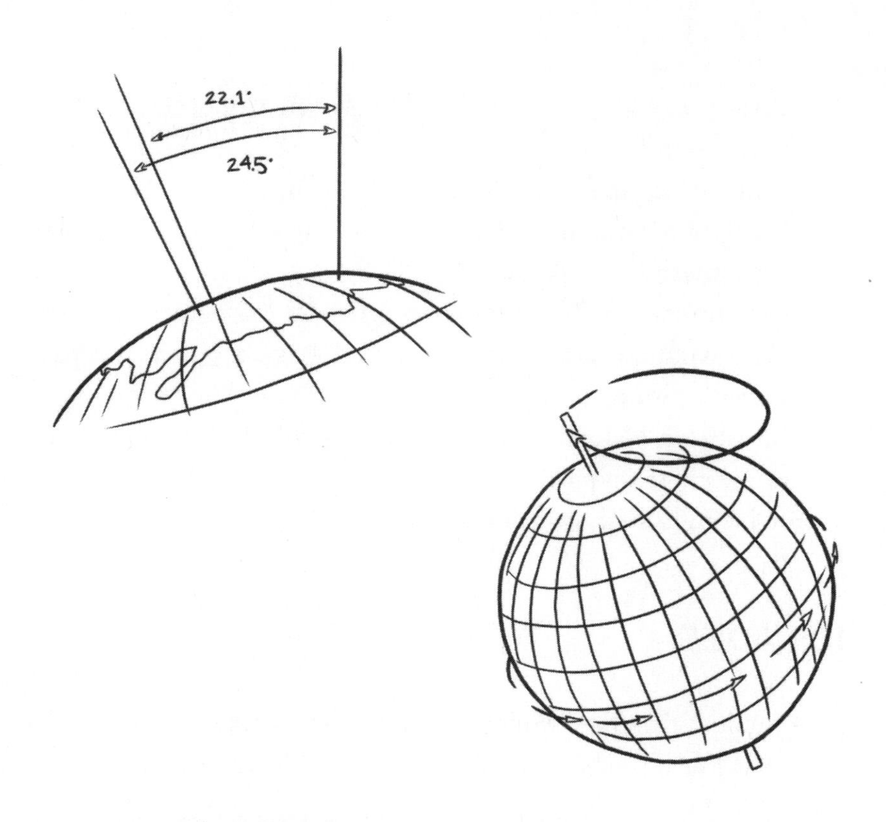

MILANKOVITCH CYCLE 3: LIKE A SPINNING TOP

The third Milankovitch Cycle that affects how much solar radiation lands on our planet is a little more tricky to understand. It's called "precession".

As our Earth orbits the Sun, the North–South spin axis does more than just rock gently between 22.1° and 24.5° on a 41,000-year cycle.

The spin axis also – very slowly, just like a giant spinning top – sweeps out a complete 360° circle. It takes about 26,000 years to do this.

So on 4 January, when the Earth is at its closest to the Sun, it's the South Pole (yep, the Antarctic) that points towards the Sun.

At the moment, everything else being equal, it's the Southern Hemisphere that will have a warmer summer because it's getting more solar radiation, but six months later it will have a colder winter.

And correspondingly, the Northern Hemisphere will have a warmer winter and a cooler summer.

But of course, "everything else" is not equal.

There's more land in the Northern Hemisphere but more ocean in the Southern Hemisphere.

The Arctic is ice that is floating on water and surrounded by land. The Antarctic is the opposite – ice that is sitting on land and surrounded by water.

The Tibetan Plateau has been rising. Big volcanic eruptions can cool the world temporarily

You begin to see how complicated it all is.

ICE AGES

As a result of the three Milankovitch Cycles interacting with each other, the Earth has had repeated short-term Ice Ages on Earth over the last 3 million years.

As a result of the three Milankovitch Cycles interacting with each other, the Earth has had repeated short-term Ice Ages on Earth over the last 3 million years.

During an Ice Age, the ice can be 3 kilometres thick and cover practically all of Canada. It can cover most of Siberia and Europe and reach almost to where London is today. Of course, the water to make this ice comes out of the ocean, and so in past Ice Ages, the ocean level has dropped by some 125 metres.

From 3 million years ago to 1 million years ago, the ice advanced and retreated on a 41,000-year cycle. But from one million years ago until the present, the ice has advanced and retreated on a 100,000-year cycle.

What we are seeing in the Arctic today – 80 per cent loss in the volume of the summer ice since 1980 – is an amazingly huge change in an amazingly short period of time.

I wonder when this loss will reach 100 per cent?

1 1.008
H
Hydrogen

Ne
Neon

29 63.546
Cu
Copper

38 87.62
Sr
Strontium

22 47.867
Ti
Titanium

Mt
Meitnerium

26.982
Al
inium

60 144.24
Nd
Neodymium

47 107.87
Ag
Silver

82

46
Pd

3 6.94
Li
Lithium

26 4.000
Fe
Iron

28 58.693
Ni
Nickel

78
Pt

49
In
Indium

80 200.59

06

BANK ROBBERY

You might have heard the old cliché "Crime doesn't pay". Well, that's not true. When it comes to Bank Robbery, the reality is that crime does pay – but most of the time it doesn't pay very well. You'd be better off working.

WAGES OF SIN

In the movies, Bank Robbery looks pretty exciting – and well paid. The hard yards consist of only a few hours of planning, followed by a few minutes of adrenaline-charged fun. And then you retire with your Fabulous Ill-Gotten Fortune to some tropical island tax haven that (rather conveniently) does not have an extradition agreement with your country – and then happily live on a diet of little drinkies with umbrellas in them.

Law, Morality and Statistics

Why doesn't everybody rob banks, if it is so glamorous and profitable? There are at least three possible reasons – Law, Morality and Statistics.

For Law and Morality, consult your local police or religious adviser. But if you *really* want to understand a topic, Statistics is the path to Enlightenment.

What about another crime that movies sometimes glamorise – selling drugs? How much does that pay?

The book *Freakonomics* is quite clear on this. Drug dealing pays so badly that most drug dealers actually live with their mothers.

That's right, in the overwhelming majority of cases, dealing in drugs is so unprofitable that you can't afford your own place, and so you have to live with your mum. Forget about the private jet, the luxury yacht and the stable of performance cars – most drug dealers actually earn much less than the average wage.

And, likewise, the average bank robber does not get fabulously wealthy.

That's what Professor Barry Reilly from the Department of Economics at the University of Sussex and his colleagues found out. They were able to access confidential data from the British Bankers' Association for the year 2007. Out of the 10,000 or so high-street bank branches in the UK, 106 suffered some kind of attempted raid or robbery in that year. That works out to about one in every hundred bank branches. (How would you like it if your workplace had those odds for some kind of armed attack?)

From the banks' point of view, they definitely don't want to lose any money, but they are also very concerned about "trauma to customers and employees and bad publicity".

Economics Transformed

Gary S. Becker made it possible for economists to analyse bank robberies. I loved his paper "Crime and Punishment: An Economic Approach". I then discovered that the great American economist Milton Friedman called him "the greatest social scientist who has lived and worked in the last half century". Becker won the 1992 Nobel Memorial Prize for Economics. He was one of the first to marry Economics to "a wide range of human behavior and interaction, including nonmarket behavior". In 1955, his thesis dealt with Racial Discrimination. He saw Economics all around him – and he radically changed this academic field.

Before Becker, Economics dealt primarily with banking, trade, inflation, business cycles, monopolies, supply, demand, etc. But thanks to his influence, today Economics also deals with the social or human side – the impact of the Oral Contraceptive Pill, crime and punishment, marriage and divorce, fertility, rational addiction (so we tax tobacco and use plain packaging to discourage teenagers from smoking), racial discrimination, and so on. When economists discuss the importance of girls' education in developing countries in the context of population, they incorporate the finding that the single most important factor in stopping over-population is educating women.

Gary Becker redefined Economics to have a social aspect. Sadly, he died while I was writing this story, on 3 May 2014, at the age of 83.

BANK ROBBERY ECONOMICS 101

But from another point of view, you can regard Bank Robbery as an Economic Activity. It has its own Inputs and Outputs, Risks and Returns, Efficiencies, and Profits and Losses.

> The average bank robber does not get fabulously wealthy.

So looking at the economics of Bank Robbery, let's examine Output. In the UK, the average return for a bank job was about £20,000. Mind you, in one-third of cases, the bank robbers got nothing at all. But there's another unhappy Output – the possibility of getting caught and being sent to gaol. This happens about 20 per cent of the time.

But what about the Inputs for the bank robbers? Inputs include, as with any economic activity, both labour and capital. These vary depending on how many robbers are in the gang, whether or not they have any firearms, and so on.

We have to factor in another economic aspect – Efficiency. How efficient are these Inputs at getting the robbers the desired Output – the big bag of money? Many factors affect the Efficiency. These include whether the bank's secret alarm is activated, how close the bank is to the nearest police station, how many customers are in the bank at the time of the raid, and so on.

BANK ROBBERY ECONOMICS – 102

It turns out that there are three main factors that affect the financial return.

First, the number of robbers. Each extra gang member raises the overall take by about £9000. Maybe this is because they are better organised. Maybe they have a more efficient division of labour. So some people could control the bank lobby, while others could carry the money out of the vault, a dedicated wheelman could be waiting to drive the getaway car, and so on. Unfortunately, the statistics tell us that with a larger crew, even though the amount of stolen money goes up, the take-home "pay" per person goes down.

Second – did the bank robbers carry weapons? Having the bank robbers armed increases the booty by about £10,000. But there's a disadvantage. Suddenly the stakes are higher, because the penalties for armed robbery are significantly greater than for an unarmed robbery.

> After one and a half years, or three bank robberies, the odds of being out of gaol are 0.8 × 0.8 × 0.8, which is roughly equal to 0.5.

The third factor is whether the bank has fast-rising security screens. These are typically powered by compressed air, and can pop up within half a second. On average, and other things being equal, the presence of these screens will reduce the size of the booty by £24,000.

ROBBERS PAY BANK!?

You might remember that I said the average take from a Bank Robbery was £20,000. But I've also just said that if the bank has fast-rising security screens, the take will be $24,000 less.

Think about it. Get £20,000, then subtract £24,000. That equals minus £4000!

Do the robbers stroll casually into a bank, notice the fast-rising security screens, extract £4000 from their wallets, hand it over to the surprised bank staff, and stroll out to instead have a nice cup of tea across the road?

Nope. That's why I (directly quoting Professor Reilly) wrote the magic words, "On average, and other things being equal . . ."

First, the statistics say the average take was £20,000. But the take ranged from zero up to £425,000. So most robberies achieved only a small take, but a tiny minority got lots of moolah.

Second, one-third of robberies got nothing. That's a lot of effort for a zero return.

Third, only about 12 per cent of British bank branches had fast-rising security screens at the time.

Fourth, you can see it's getting messy. Do the Maths and combine factors one, two and three.

Professor Reilly did this and summarised the analysis. He wrote that about two-thirds of that £24,000 reduction was due to the lower chance of a raid being successful, and about one-third because they took away much less money.

But let's get back to the average Bank Robbery of £20,000. It typically involved 1.6 bank robbers. So the return was only about £13,000 per bank robber. That's about half the average UK full-time wage of around £26,000. So our bank robbers would not be set for a life of wanton luxury in the Bahamas. All they get is a modest lifestyle for half a year.

But associated with each Bank Robbery is an 80 per cent chance of not being caught. After one and a half years, or three bank robberies, the odds of being out of gaol are 0.8 × 0.8 × 0.8, which is roughly equal to 0.5. In plain English, after three Bank Robberies, the chance of getting caught by the cops and ending up in gaol is 50 per cent. After four robberies, there's a 60 per cent chance of getting caught. As a career path and an economic activity, robbing banks is rubbish.

The lessons are obvious. Economic Rationalism is quite clear on this. Preferably, go straight – or if you can't, change over to White Collar Crime . . .

Dumb and Dumber

In 2005, two 19-year-old Antipodeans (Luke Carroll and Anthony Prince) bumbled their way through an armed Bank Robbery in the skiing town of Vail, Colorado, in the USA.

They each wore their name tags from the sporting goods store at which they worked. Furthermore, they were regular customers at the bank they robbed, and did not attempt to disguise their distinctive Australian accents. Not surprisingly, they were easily identified.

They stole more than US$130,000, mostly in low denominations. The cash was so bulky that they couldn't carry it all, so they dumped $2000 in $1 bills.

They then realised the cash was too bulky to travel with. They tried mailing some home. Anthony Prince tried to buy an expensive Rolex watch with $5 bills. They gave $20,000 to a cab driver. At the airport, they tried to buy one-way tickets to Mexico with cash.

Yes, they were arrested at the airport.
Carroll served 4.5 years in prison,
while Prince served 5 years.

07

CALORIE IS NOT A CALORIE

Great Nobel Laureate Richard Feynman, said something very wise that ran along the lines of, "Science is a way to not get fooled." So you would hope that once you understood a subject so deeply that you could start using numbers to describe it, you would be well on the way to The Truth.

But what if the numbers themselves were "confusing"?

As an example, in the land of Weight Loss, it turns out that a calorie is "less" than a calorie!

Yes, if you eat a single calorie, by the time it has been processed and made available for your body to use, it will be equal to less than one calorie. Sometimes, it's a little bit less, but sometimes it's a lot less. (This seems like a weight-loss dream scenario – but it's true.)

A CALORIE IS LESS THAN A CALORIE?

Sure, if you are overweight, reducing how many calories you eat is an important part of achieving and maintaining a healthy weight.

But how do we begin to count calories?

Let's ask Wilbur Olin Atwater. He set up a system, in around the year 1900, to work out the number of calories assigned to each food.

He simply "burnt" each food type (fat, protein and carbohydrate) in a machine called a Bomb Calorimeter and measured the temperature rise. He then applied basic physics to calculate how much energy had been generated. He used this to work out how much energy was provided by the individual macronutrients of fat, protein and carbohydrate. We still use this system today. It works, but it's a little crude.

It turns out that the actual energy delivery in the body is more complicated than can be calculated in a machine.

Yes, 100 calories of any food will generate the same amount of energy if you burn it in a Bomb Calorimeter. But it will generate fewer calories for you once inside the human body which is very complex. You have to allow for hormones, gut bacteria, genetic makeup, metabolic rate, and many other factors.

"Calorie" or "Kilojoule"

In the Good Old Days, "energy" was measured in "calories". For various reasons, we changed over from "calories" to "kilojoules".

Just for reference, one calorie is about 4.2 kilojoules.

The problem is that calories have been around for ages and are so well known. People don't like to change from what they're used to.

We have a similar problem with the weights of babies. Even though Australia went metric in 1976, many newborn babies are still weighed in pounds, not kilograms.

By a lovely coincidence, a kilojoule is roughly the amount of energy that the Sun radiates onto one square metre of the surface of the Earth in one second – when the Sun is directly overhead.

ENERGY PROCESSING TAX

First, it costs your body different amounts of energy to digest different foods.

Food has to be broken down or processed into very small molecules. Only small molecules can cross the gut wall. The Dieticians call the energy needed to do this processing and transfer the "Thermic Effect of Food" or "Energy Tax".

Fats use only 2 to 3 per cent of their embodied energy in this breaking-down process, so they have the lowest Energy Tax. (That leaves 97 to 98 per cent of their potential energy available to your body.) Carbohydrates are a bit higher at 5 to 10 per cent. But proteins use a massive 25 per cent of their energy, mostly to unravel the tightly wound amino acids that make them up.

So if you eat 100 calories of fat or 100 calories of protein, your body will ultimately have access to 98 calories for the fat, but only 75 calories for the protein.

(By the way, each different variety of fat, protein and carbohydrate will have a different Energy Tax. I have given approximate values.)

Chocolate Has Negative Calories?

I was initially delighted to read a study that looked at the link between chocolate consumption and fatness in 1458 European adolescents. If they ate a lot of chocolate, they were less fat.

I experienced great jubilation. For a moment I dared wonder: does this mean that the calories in chocolate are special, and that they actually suck fat out of your body?

Unfortunately, no.

It turns out that this was a poor study.

First, the adolescents self-reported how much chocolate they ate. Self-reporting has very many unintentional measurement errors. Second, adolescents (12.5 to 17.5 years of age) are notorious for misreporting. Third, the dietary intake was measured on only two days – hardly a reasonable sample size.

And fourth, Correlation does not equal Causation. There is a close correlation between "Age of Miss America" (ranging from 19 to 24 years) and "Murders by Steam, Hot Vapours and Hot Objects in the USA" (ranging from two to eight per year). Did one cause the other? No.

But I do still wish that chocolate had negative calories.

YES, A CALORIE IN FOOD IS DEFINITELY MORE THAN A CALORIE IN YOUR BODY

Second, the Atwater method of calculating calories deals with the total potential amount of energy in the food. But, in some foods a significant amount of the energy does not get extracted, but goes straight through your gut into your faeces.

> If you eat a single calorie, by the time it has been processed and made available for your body to use, it will be equal to less than one calorie.

Almonds might deliver 170 calories of energy in a Calorimeter. But once inside your gut, 41 of those calories go straight through to the toilet bowl, leaving only 129 potentially available to you. As another example, seeds can pass through your gut so unscathed that, if planted, they will sprout and grow. Once again, fewer calories are available to your body.

Food preparation is a very important third factor in the Calorie Is Not a Calorie story. Depending on the food, you can make a lot more of its energy available to your body, if you process it first. Some options include grinding or pounding it to increase its surface area, cooking it (boiling, baking, grilling, sautéing, etc.) or fermenting it.

For example, mice actually lose weight when they eat only raw sweet potato. However, they gain weight when eating the same weight of cooked sweet potato. In this case, the heat breaks down the very tough cell walls of the sweet potato, making the contents more easily accessible.

With regard to meat, rats gain a little weight when eating raw meat, but gain more weight after eating the same weight of cooked meat.

A fourth (and rather major) factor is that people are different from one another.

For example, two-thirds of humans are Lactose Intolerant to some degree. (See story on page 151.) If you are very Lactose Intolerant, when you drink milk, you will lose weight – thanks to diarrhoea. But if you are Lactose Tolerant, you will gain weight.

> If you eat 100 calories of fat or 100 calories of protein, your body will ultimately have access to 98 calories for the fat, but only 75 calories for the protein.

Another difference between humans is the length of our gut. Back in the early 1900s, some European scientists decided to measure the length of the intestine. They discovered that a few specific Russian populations had an extra half-metre of bowel compared to some Polish populations. The longer gut meant that these Russians could extract more energy from their diet than could the Poles.

And another variable is that we all have different populations of microbes in our digestive tracts. This is the so-called "gut microbiome". If your gut has been colonised by the *Firmicutes* bacteria, they will provide you with extra amounts of energy – and you will gain weight. The *Bacteroidetes* bacteria have the opposite effect – and you will lose weight. (This is discussed in painstaking yet entertaining detail in a previous book of mine, *Brain Food.*)

And finally, it costs us more energy to digest whole foods than processed foods. Compare two cheese sandwiches, each rated at around 700 calories. One is whole-wheat bread with sunflower seeds, kernels of grain and cheddar cheese. The other is white bread with "processed cheese product". The people who eat the whole-wheat bread sandwich obtain 10 per cent fewer calories.

So what goes around doesn't always come around . . .

Sorry, "Calorie" = "1000 calories"

I'm sorry to have to tell you this, but the word "calorie" has two completely different numerical values. One is a thousand times bigger than the other.

If you talk to a Physicist, a "calorie" is the amount of energy needed to increase the temperature of a single gram of water by a single Celsius degree.

But if you talk to a Dietician, a "calorie" is the amount of energy needed to increase the temperature of 1000 grams of water by a single Celsius degree. Yup, it's the same word, but its size is 1000 times bigger. If you push the Dietician, they will admit that they also call it a "large calorie", a "kilogram calorie", a "dietary calorie", a "nutritionist's calorie", a "nutritional calorie", and even a "Calorie" (note the Capital letter at the beginning of the word). Probably the best of all is a "food calorie".

So a "small calorie" equals about 4.2 joules, but a "food calorie" equals about 4.2 kilojoules (or 4200 joules).

By the way, the word "calorie" comes from the Latin *calor*, which means "heat".

08

CIGARETTES: ROLL YOUR OWN VERSUS FACTORY MADE

Back when I was in my 20s, I foolishly took a substance that today is still the most dangerous and addictive drug known to the human race. This psychoactive substance kills more people than any other drug. Yup, for a few years, I stupidly smoked cigarettes.

At the time, I was under the delusion that by smoking Roll Your Own cigarettes, or Rollies, rather than Factory Made, I was taking a healthier (or, at least, less harmful) pathway. I was so wrong. Rollies contain about 20 times more additives than Factory Made cigarettes.

HEALTH COSTS OF TOBACCO

Tobacco is the most widely used drug of abuse on the planet. It is also highly addictive. Each year, about half of all smokers attempt to quit – but only a small percentage are successful in giving up long-term.

Tobacco causes 89 per cent of all drug-related deaths. Averaged out over the population, it is responsible for 11 per cent of deaths from all causes.

Half of the people who smoke cigarettes long-term will die from a smoking-related disease.

In Australia, that works out to about 15,000 early deaths per year. None of these premature deaths had to happen. Smoking kills you mostly by various cancers, heart disease, and chronic obstructive lung disease – plus many other pathways.

Cancers are the number one cause of smoking-related deaths, killing just over half of male and female smokers. About three-quarters of these cancers are lung cancer. In fact, lung cancers cause most of the cancer deaths in Australia – overwhelmingly due to smoking.

To summarise, tobacco causes about 89 per cent of all drug-related deaths, and about 11 per cent of all deaths – and will kill half of its long-term smokers. Furthermore, it costs the taxpayer money. But Big Tobacco benefits.

Financial Costs of Tobacco

The financial costs of tobacco are enormous.
In Australia, these monetary costs are about six times bigger than the tax revenue that comes from cigarette sales.

In the Australian Financial Year of 2004/2005, smokers spent over 750,000 bed-days in hospital – leading to hospital costs of about $670 million. But hospital bed-day costs are just the tip of the iceberg.

When you factor in *all* the health costs, they add up to about $31 billion each year. This $31 billion dwarfs the revenue gathered from taxing tobacco, which is only about $6 billion per year.

(Minor economic note: taxes are not an economic cost or benefit. They are a transfer payment. The government gets richer, while the taxpayer – the smoker – gets poorer.)

Mind you, apologists for Big Tobacco like to ignore this basic financial fact. They try to muddy the waters by wrongly claiming that that cigarettes actually "save" the government money. This supposed "saving" comes from people dying sooner – and so not drawing on the pension.

But Big Tobacco says nothing about lost productivity, costs of prescriptions, help from ancillary services, medical and hospital visits, chemotherapy and radiotherapy costs, nursing-home care, fires attributable to smoking, ambulance transport, and so on and so on. The children of smokers are four times more likely to be admitted to hospital for a respiratory condition. Even before smokers slide down their final pathway, they spend more unproductive time at work, take more sick leave, and use 20 per cent more hospital days than non-smokers.

Even the smokers themselves suffer financially. Heavy smokers have "a reduction in net worth of almost $8400, compared to non-smokers" per year. And "financial stress [is] more frequent in households where a smoker lives, regardless of the household income".

Tobacco robs the Australian economy of about $25 billion each year. The health costs far outweigh the revenue.

ROLLIES – WHO AND WHY

Most of us are familiar with the neat cylindrical sticks of Factory Made cigarettes in their cute little rectangular cardboard packs. But Rollies, which are mostly hand-rolled with loose fine-cut tobacco and cigarette papers, are usually fairly messy. You can make them slightly less messy if you use a little portable rolling machine.

Rollies have about 20 times more additives than Factory Made cigarettes.

The smoking of Rollies rather than Factory Made cigarettes varies from country to country. The popularity of Rollies in the smoking population ranges from about 58 per cent in Thailand, 39 per cent in New Zealand, 28 per cent in the UK, 24 per cent in Australia, 17 per cent in Canada, right down to only 7 per cent in the USA. On average, it's poorer people and heavier smokers who smoke Rollies more frequently. They usually do not intend to give up smoking, and have more friends who smoke.

WHY DO PEOPLE SMOKE ROLLIES?

Again, it varies around the world. But in general, there are three main reasons. First, they're cheaper. Indeed, if you're feeling especially broke, you can roll really thin cigarettes, and use less tobacco. Second, Rollie smokers feel more independent and "cool" and part of a special sub-culture. But third is the belief that Roll Your Own tobacco is more "natural", and therefore "safer" or "less harmful" than Factory Made cigarettes.

Unfortunately for the people who smoke Rollies, these cigarettes are both less natural and more harmful.

ADDITIVES – 20 TIMES MORE

In Factory Made cigarettes, the additives make up about 0.6 per cent of the dry weight of the tobacco. But in Rollies, the additives weigh in at a colossal 13 per cent of the dry weight. (The figure is about 17 per cent for pipe tobacco.) In other words, Rollies have about 20 times more additives than Factory Made cigarettes.

These additives include sweeteners, such as honey, sugar or dextrose. A cynical person might think that these sweeteners make Rollies "smoother" for new smokers, such as children.

But some of the other additives include chemicals with less-than-wholesome names such as propylene glycol, glycerol, trans-benzaldehyde, phenylcarbinol and ethyl butyrate. Another factor in the health risk of Rollies is that they are usually thinner. This means that you have to suck harder, which increases the amount of tar that lands in your airways.

MORE THAN NICOTINE

Tobacco smoke carries nicotine – and over 4000 other chemicals. We still don't fully understand what these other 4000+ chemicals do to the human brain, and the body as a whole.

A Norwegian study of 26,000 smokers showed that Rollies were associated with higher, not lower, risks of lung cancer. And other studies consistently show a 2 to 3 times increased risk of cancers of the oesophagus, mouth, pharynx and larynx directly linked to Rollies.

We don't know why.

Could it be that most Rollies do not have a filter? Or is it because Rollie smokers "may take more puffs, inhale more smoke per cigarette, and for longer periods, and have been reported to be less likely to make quit attempts than those who smoke factory-made cigarettes"? Or is it because of the 21 times more additives than Factory Made cigarettes?

The way that the tobacco companies get around the increased health risks of Rollies is by totally ignoring the facts – and "bending" the truth. Big Tobacco is now changing the image of Rollies from low-cost and down-at-heel to "cool", "better tasting" and "natural and organic".

Big Tobacco Disinformation

Since the 1950s, the tobacco companies have run consistent, massive and well-funded disinformation campaigns. Mostly using proxies and apologists, they use selective reporting, distortions of the truth, half-lies and outright lies. These "third parties" concentrate mainly on the Health Issues, but are quite happy to do the same for Financial Issues.

Do you want to know the fine detail of how Big Tobacco does its dirty work? Read *Merchants of Doubt: How a Handful of Scientists Obscured the Truth on Issues from Tobacco Smoke to Global Warming* by Naomi Oreskes and Erik M. Conway.

THE GOOD NEWS

There is potential good news – no matter what kind of cigarettes you smoke. Your health can recover, once you stop smoking.

If you quit smoking before middle age, you reduce the risk of lung cancer by 90 per cent. Within two to five years of giving up smoking, your risk of heart attack and stroke drops significantly.

After 15 years of not smoking, your risk of stroke is the same as if you had never smoked.

ILLOGICAL INHERENT CONTRADICTION

All this ignores an Inherent Contradiction – or, in Plain English, "something that doesn't make sense".

On one hand, overwhelmingly around the world, it is official government policy to discourage smoking. But on the other hand, we allow companies (that exist only to make a profit) to sell dangerous addictive drugs to the public which directly kill half of those who get addicted.

Why can't we just tell the cigarette companies to simply butt out?

09

DANCING WITH DEER EVOLVES YOUR BRAIN

As a percentage of our weight, we humans have the biggest brains in the Animal Kingdom. But looking at the rest of our anatomy, we seem to be nothing special. However, we do have one physical "gift". Surprisingly, it turns out that a well-trained, fit human can outrun a deer.

Over the past few million years, we evolved towards modern-day you-and-me *Homo sapiens*. We also evolved the ability to be good endurance runners. As often happens with evolution, there was a "benefit". It was that we could chase game animals for hours, run them to a standstill, kill and then eat them. A current theory in Evolutionary Biology claims that eating the concentrated protein and energy of game animals helped our brains evolve bigger.

In other words, our ability to run long distances and our ever-increasing brain evolved together – one helping the other.

Walking versus Running

All the Exercise Physiologists I know are quite lean. They all run. Many have run a Marathon (42.195 kilometres).

Running is quite different from the sport of Racewalking. A Racewalker has to walk as quickly as possible without actually breaking into a run, while always keeping one foot in contact with the ground. Top Racewalkers can reach around 15 to 16 kilometres per hour.

Marathon runners are a little scornful of Racewalkers. They describe Racewalking as trying to whisper as loudly as possibly, without actually speaking.

MAN OUTRUNS DEER

Back in 1978, Michael Baughman wrote an article in *Sports Illustrated* about how he outran a deer. His great-grandfather had done this in his youth. He had also told Michael how their Mohawk Indian ancestors had run down deer for food.

The deer is more of a sprinter than a marathon runner. On a warmish day (around 27°C) it took him about four hours to run down his chosen deer over a distance of 24 kilometres. The countryside was near his home – open rangelands with fairly gentle hills, and occasional willow thickets. The vegetation was sparse enough that he could always see for a few hundred metres, so the deer couldn't vanish from his sight – and recover its strength while hiding. Michael's average speed was 6 kilometres per hour. This is not much faster than a quick walking pace, but the surface wasn't a flat paved road.

After four hours of chasing, the deer could not run or even walk – it had nothing left in the tank. Michael sidled up to within a few metres of the exhausted animal, all the time talking quietly and soothingly, and gently touched the deer's sweaty flank. And then, unlike our primeval ancestors, he let the deer escape.

Why could Michael run down the deer? Not just because he had trained to run a marathon in under three hours – there was also some basic Biomechanical Physiology involved.

COST OF TRANSPORT

Let me introduce you to "COT" – the metabolic Cost of Transport. It's how much oxygen you have to breathe in to shift a kilogram of body weight a distance of one kilometre. For most animals, it's a U-shaped curve. The bottom of the U is where you use the least amount of energy.

A horse has three of these U-shaped overlapping curves. (Dr Bramble and Dr Lieberman, who researched this, ignored the "canter", regarding it as a low-speed gallop.) For walking, the horse's most efficient speed is around 4.3 kilometres per hour, for trotting it's around 11.9 kilometres per hour, while galloping is most efficient at around 22.7 kilometres per hour. The horse will transition from walking to trotting to galloping at the points where these curves cross over.

For the average human, the most efficient walking speed is around 5.4 kilometres per hour (it varies with the person). Walking at speeds either slower or faster than 5.4 kilometres per hour is less efficient. We burn up more oxygen and energy to shift our weight over a given distance.

But as our walking speed increases to around 7.9 kilometres per hour, the two COT curves for "walking" and "running" cross.

Travelling faster than 7.9 kilometres per hour, we actually burn less energy and oxygen by running instead of walking.

This is because the act of running uses what the physicists call a "Mass-Spring Mechanism". Various tendons and ligaments in our legs act as springs to capture and store the elastic strain energy as our legs hit the ground. They then release this energy through recoil, when we push off onto the next step.

Mechanism of Walking

The first time I pulled a wheelie bag inside an airport, I noticed that I was constantly speeding up and slowing down with each step. This was the first time I realised I did not walk at a constant speed. I could feel the handle on the palm of my hand. For one part of the step the handle was dragging behind me – but for another part it was pushing on my hand.

I got more proof when I pulled the wheelie bag over a rough surface. I could easily hear the wheels speed up and slow down – with each step.

Without even thinking about it, I was swapping Height for Speed with each step.

Let me explain.

As I walked, my Centre of Mass bobbed up and down with each step. It was highest in mid-step when my two legs were straight and directly under me. My Centre of Mass was lower when my heel struck the floor, and when I pushed off on my toe. So as I walked, I unthinkingly swapped Potential

and Kinetic Energy back and forth with each step. (The physicists call this system an "Inverted Pendulum".)

When my Centre of Mass was highest, I was slowest – and vice versa.

The Mechanism of Running is quite different. When we run, we store energy in biological "springs".

THE EVOLUTION OF RUNNING – FOUR CHANGES

How did we get to be such efficient long-distance runners? After all, none of the other primates (two-legged animals with hands, hand-like feet and forward-facing eyes) can run.

For the average human, the most efficient walking speed is around 5.4 kilometres per hour

Evidence shows that the long process of evolution began with our ancient primate ancestors. We have found strong hints of these evolutionary changes in *Homo habilis* some 2.6 million years ago, in *Homo erectus* 1.8 million years ago and in *Homo ergaster* some 1.6 million years ago.

Endurance running imposed four major demands on our evolving bodies – Energetics, Strength, Stabilisation and Thermoregulation.

The solution for Energetics is the Mass-Spring Mechanism. In the human, the main springs are the Achilles tendon and the longitudinal arch of the foot. In fact, when we run, the arch of the foot just by itself returns about 17 per cent of the energy captured in each step. Another energy factor is Stride Length. We humans differ from the four-legged animals by increasing our speed mostly by increasing the length of

Between 2 million and 200,000 years ago, our brain evolved from 600 cubic centimetres to its current volume of around 1300 cubic centimetres.

our stride, rather than having more strides per minute.

The second factor needed for endurance running is having a skeleton strengthened in the relevant critical areas. For example, compared to the other primates, we humans have a rather sturdy femoral neck (the top of the leg bone, where it slots into the hip joint). This means that the femoral neck doesn't bend as much when our foot slams into the ground. Also, most of our leg joints are wider than in the other primates. This gives a bigger surface area across which to spread the load. Our toes are shorter – again, an advantage in running.

Stability is critical for a two-legged animal. After all, walking is a series of "controlled falls". The act of running is even worse, because we have less time to straighten up after each "fall". The single leg that touches the ground generates enormous twisting forces as it pushes off. So we evolved features to make our trunk more stable – such as greatly enlarged muscles on the buttocks and on the spine. There are also changes in the neck and skull to enhance stabilisation of the head (so that it doesn't whip around). None of the other primates have these changes. We are the only primate that can run.

Finally, Thermoregulation. With regards to getting rid of heat, we humans have a massive advantage over all the other primates, and most of the quadrupeds. We have a superior ability to get rid of the excess metabolic heat generated by running. We have huge numbers of sweat glands. As sweat evaporates, it takes with it an enormous amount of heat. We also have less body hair, and a narrow, long body to radiate away the heat. Our very efficient mouth-breathing gives us higher airflow rates (with lower muscular effort), as well as another pathway to dump excess heat. Other primates can't do long-term deep mouth-breathing.

FOOD OR FUN

So we are much more efficient at moving quickly than the other primates.

But even with these four evolutionary changes, we humans are not as efficient as other four-legged mammals. Weight for weight, we still burn twice as much energy to cover a given distance at speed. But back before we developed agriculture, we were running to get our next meal, not just for fun. (I wrote about this transition to Agriculture and how it initially reduced our life span but increased our population size enormously in my 17th book, *Flying Lasers, Robofish and Cities of Slime.*)

Endurance Runners

Today, we humans are lousy sprinters. Usain Bolt, the Olympic sprinter, has briefly reached 45 kilometres per hour – but only during a 10-second sprint. However, endurance is another story. Ultra-marathon runners have run 270 kilometres in just 24 hours.

Only a few other mammals can do endurance running. These include the social carnivores (dogs, hyenas, etc.) and migratory hoofed animals (horses, wildebeest, etc.).

The theory is that as we evolved endurance running, we could also push our prey to exhaustion, and then kill and eat them. It turned out to be good for our brain. Between 2 million and 200,000 years ago, our brain evolved from 600 cubic centimetres to its current volume of around 1300 cubic centimetres.

But today, the furthest that the majority of us have to run to chase our dinner is the few steps to the fridge. If we had to resort to running down a deer to get our evening meal, most of us would run into trouble . . .

Chasing Bunnies and Kangas

On my ABC homepage, Chris wrote a comment, "We used to regularly run down rabbits. Just keep them running/moving away from refuge (burrows, scrubs, etc.) for a few hundred metres and they fatigue very quickly." I was surprised.

I was even more surprised by Ryan's comment, "Ask any kid growing up on a farm – running down a kangaroo is incredibly easy. Takes about four k's and an active 12-year-old can handle the task without a problem. The kangaroo will tire even quicker if the ground is uneven or scrubby."

10

EAT LESS, MOVE MORE

People in the world's wealthier countries are getting fatter. In the USA, about two-thirds of the population are overweight, and about one-fifth are obese.

This Obesity Epidemic has happened despite an enormous increase in the number of weight-loss diets and low-fat foods available. The most recent weight-loss research seems to show that diets don't have to be so complicated.

If your goal is long-term weight loss, the actual diet you follow doesn't matter. It's irrelevant whether you reduce just fats, or just proteins or just carbohydrates. The simple message is that you have to reduce your total calorie intake, and do more exercise.

EAT LESS, MOVE MORE

In one typical two-year-long study, the subjects both reduced their calorie input and did some 90 minutes of extra exercise each week. The goal was to reduce their weight.

Overall, the group's health risk factors improved. HDL (or "good") cholesterol increased, while LDL (or "bad") cholesterol decreased. Triglycerides, insulin and blood pressure all dropped. The long-term average weight loss was about 6 kilograms. The dietary ratios of fats, proteins and carbohydrates were largely irrelevant.

As a result of the study, the advice of the authors was to choose a diet that was balanced, satisfying and did not leave you hungry. They found that foods with a high Glycaemic Index tend to spike your blood sugar level, shooting it straight up, and then straight down again to below normal levels. This then stimulates your hunger. (Glycaemic Index is simply explained in a previous book of mine, *Brain Food*.) If your goal is to lose weight, you should avoid high Glycaemic Index foods.

The subjects did better with their weight-loss goals if they had regular sessions with counsellors. So behaviour modification can also help.

FAT, PROTEIN OR CARBOHYDRATE?

A lot of diets tend to focus on the macronutrients of our food – the fats, proteins and carbohydrates.

People (sometimes unscrupulous, sometimes well-intentioned) have made buckets of money by writing books demonising, or promoting, particular macronutrients.

Some diet books might say eat only protein and fat – but never, ever, ever eat carbohydrates. Other diet books might claim that proteins are "evil". Sometimes the claim is that a particular combination is the problem. And so on.

By the way, very few of these authors are qualified dieticians.

Obviously, the various authors can't all be correct.

What if it turns out that none of them are correct?

Dietician or Nutritionist?

In general, the word "Dietician" is "protected". To call yourself a Dietician, you usually have to do a University Degree.

But anybody can call themself a "Nutritionist" – even if they have zero training.

THE "SIMPLE" ANSWER

While actually losing weight might not be simple, weight-loss plans don't have to be complicated.

> The simple message is that you have to reduce your total calorie intake, and do more exercise.

Part of the problem is that we have 200,000 years of evolutionary history, during which we did not have the moderately secure food supply from agriculture that most of us in wealthier countries enjoy now. The evolutionary imperative was straightforward – if food is there, eat it right now, because it might not be there tomorrow.

Fortunately, research (the stuff in the peer-reviewed scientific literature, not the latest Diet Fad) can help.

Research has found that the differing proportions of fat, protein and carbohydrate in your diet don't really matter in terms of long-term weight loss.

To lose weight, you have to reduce overall calorie intake in a way that is healthy for your heart. In other words, eat smaller amounts of saturated fats and cholesterol, and greater amounts of dietary fibre. The individual macronutrient choice (fat, protein or carbohydrate) doesn't matter.

The highly consistent factors in successful weight-loss diets were reducing calories and doing a bit more exercise. For example, get a dog and take it for walks.

This is the weighty question: how do we make a simple message about eating less and moving more easier to digest . . .

11 THE FIVE-SECOND RULE

An enduring Food Myth is the Five-Second Rule (occasionally called the Three-Second Rule). This rule claims that it's okay to eat food that you've dropped and picked up from the floor – so long as you pick it up off the floor within that magical time of Five (or Three) Seconds. This myth might have stood the Test of Time, but it doesn't pass the Test of Truth.

So let's just imagine the tasty food morsel dropping onto the floor. Are we supposed to believe that the bacteria all start their stopwatches and hold themselves back, waiting a whole five seconds before they leap onto the Foody Offering from the Gods above? No, they dive in.

Food Poisoning

Food Poisoning is usually more inconvenient than life-threatening. But there is a dark side.

In the USA, each year there are some 76 million cases of Food Poisoning. Of these people, some 300,000 are hospitalised. But 5200 don't get better – they die from Food Poisoning.

The UK figures are about one million cases per year, 20,000 people hospitalised and up to 500 deaths.

RESEARCH – PART 1

One early piece of research into the Five-Second Rule was done in 2003, by Jillian Clarke. She was a student from the Chicago High School for Agricultural Sciences, doing a seven-week summer apprenticeship at the University of Illinois. This was an intensive laboratory and academic program for talented high school students.

> The *E. coli* bacterium transferred from the tile to the food in five seconds or less.

For her ground-breaking research, she was awarded the 2004 Ig Nobel Prize in Public Health. (Mine was for Belly Button Fluff – just saying . . . Read more in my 20th book, *Q&A with Dr K.*)

Ms Clarke started with a survey. Some 56 per cent of men and 70 per cent of women knew of the Five-Second Rule. It turned out that women were more likely than men to pick up and eat food from the floor.

Not surprisingly, she also found that biscuits and lollies ("cookies" and "candy", as they call them in the USA) were much more likely to be picked up and eaten than broccoli or cauliflower. (As it turns out, "natural" foods such as vegetables, meat and cheese have higher levels of naturally occurring micro-organisms. Maybe that's why they weren't picked up and eaten as often as the "cleaner" biscuits and lollies?)

Then Ms Clarke did some experiments. There were many factors to consider. You might compare it to touching something hot, such as a cast iron frypan. What are the potential injuries, and how severe might they be? There are many factors involved – how hot it is, how long you touch it for, which part of it you touch, and whether you grab it with your whole hand or it just brushes against your clothing.

Ms Clarke coated the surfaces of rough and smooth tiles with the bacterium *E. coli*, and then placed either Gummy Bears or fudge-striped cookies on them.

In all cases, the *E. coli* bacterium transferred from the tile to the food in five seconds or less. More bacteria transferred across from the smooth tiles than from the rough tiles. That sounds reasonable – the smooth tiles would have more surface area to make contact with the biscuits and lollies, so more bacteria could migrate across. Indeed, later studies showed that more bacteria transferred across to wet, floppy lettuce than to dry, stiffer lettuce.

Poisoned Food

In one US study, some 70.7 per cent of the poultry carcasses and 91 per cent of the retail chicken products were contaminated with *Campylobacter* species. These are nasty bacteria.

They can be killed by cooking. But suppose the chicken was temporarily placed on a breadboard, and then transferred

to the stove for cooking. If that breadboard was not cleaned properly, and used again for preparing (say) sandwiches, some of the *Campylobacter* could transfer across to the bread.

The technical term is "bacterial cross-contamination on food preparation surfaces".

RESEARCH – PART 2

The next major research was carried out in 2006 by Professor Paul Dawson and colleagues, at Clemson University in South Carolina. Their experiments were similar to those of Ms Clarke – but more detailed.

They inoculated three surfaces – wood, tile and carpet – with a rather nasty bacterium called *Salmonella typhimurium*. Salmonella is pretty powerful – it can cause Food Poisoning with as few as just 10 individual bacteria. The researchers applied bacteria in doses of several million per square centimetre. After a day, there were thousands of bacteria per square centimetre still remaining on the tile and wood, and tens of thousands per square centimetre on the carpet.

Salmonella could survive on all these surfaces thanks to its making and residing inside a Living Goo called a "Biofilm". A Biofilm is made up of multiple layers of polysaccharides that also incorporate proteins and fats. A Biofilm not only traps nutrients for the bacteria within, but also protects them from hostile environments around them. (I wrote about Biofilms in the story "Cities of Slime", in my 17th book, *Flying Lasers, Robofish and Cities of Slime*.) Dawson's team found that the bacteria could survive for a month – and still be present in sufficient number to cause Food Poisoning.

The scientists then used bologna sausage as a food to "catch" the Salmonella. They measured how many bacteria transferred from the floor to the sausage.

The highest rate of Salmonella bacteria transferring from the Biofilm to the bologna was from the hard smooth tile – over 99 per cent of the bacteria had moved across in less than five seconds. The transfer rate was much lower with the carpet – less than 0.05 per cent of the bacteria made it across. That makes sense because there was very poor contact between the individual strands of carpet and the sausage resting on them.

We are surrounded by bacteria – internally and externally. They make up some 90 per cent of the cells in our bodies.

"Protective" Food

It turns out that the rate at which the bacteria transfer to the food depends, to some degree, upon the food itself. Food that is high in salt or sugar (as compared to healthier foods with more water, such as fruit) forms a hostile environment inside the food. As a result, the food takes up bacteria at a lower rate.

By the way, this is *not* an express directive to eat junk food. Rather, try not to drop food and then eat it – even if it is lollies.

RESEARCH – PART 3

The most recent research was carried out in 2014 by Professor Anthony Hilton and six final-year biology students at the School of Life and Health Sciences at Aston University in the UK. It has not yet been published in a peer-reviewed journal.

They found that (as expected) bacteria are less likely to transfer from carpet, and more likely from wood or tile.

They explored the effect of time. Bacteria certainly transfer across immediately. But increasing the time by a factor of 10 (from 3 to 30 seconds) increased the number of bacteria that migrated by up to 10 times.

Professor Hilton did emphasise that there is "a risk inherent in eating anything from any type of floor".

POO ON YOUR COFFEE TABLE?

But there's another complication. Most of us wear our street shoes into the house.

A study by Charles Gerba, Professor of Microbiology at the University of Arizona, found that after three months of use, 93 per cent of shoes have faecal contamination. After all, you walk into private and public bathrooms, and on grass and gutters, wearing your shoes.

Even more disconcerting, he studied coffee tables in the apartments of single men. About seven out of ten of these coffee tables had faecal contamination. This almost certainly happened because the blokes put their still-shod feet up on the coffee tables after a hard day's work.

(Mind you, a little skepticism might be in order. This study was funded by Clorox, which makes Disinfecting Sprays, Germicidal Bleaches and Anywhere Hard Surface Daily Sanitising Spray.)

THE ZERO-SECOND RULE

We are surrounded by bacteria – internally and externally. They make up some 90 per cent of the cells in our bodies, and they are on all the surfaces around us. In the majority of cases, we humans and bacteria exist in harmony.

People do talk about the so-called "Hygiene Hypothesis" – that we are "too clean" and need to be exposed to more germs in the environment to strengthen our immune systems. But eating food that has dropped onto the floor is not the right way. The Five-Second Rule should be the Zero-Second Rule. When in doubt, throw it out.

If you want to build up your immune system, eat and sleep well, do exercise – and get vaccinated.

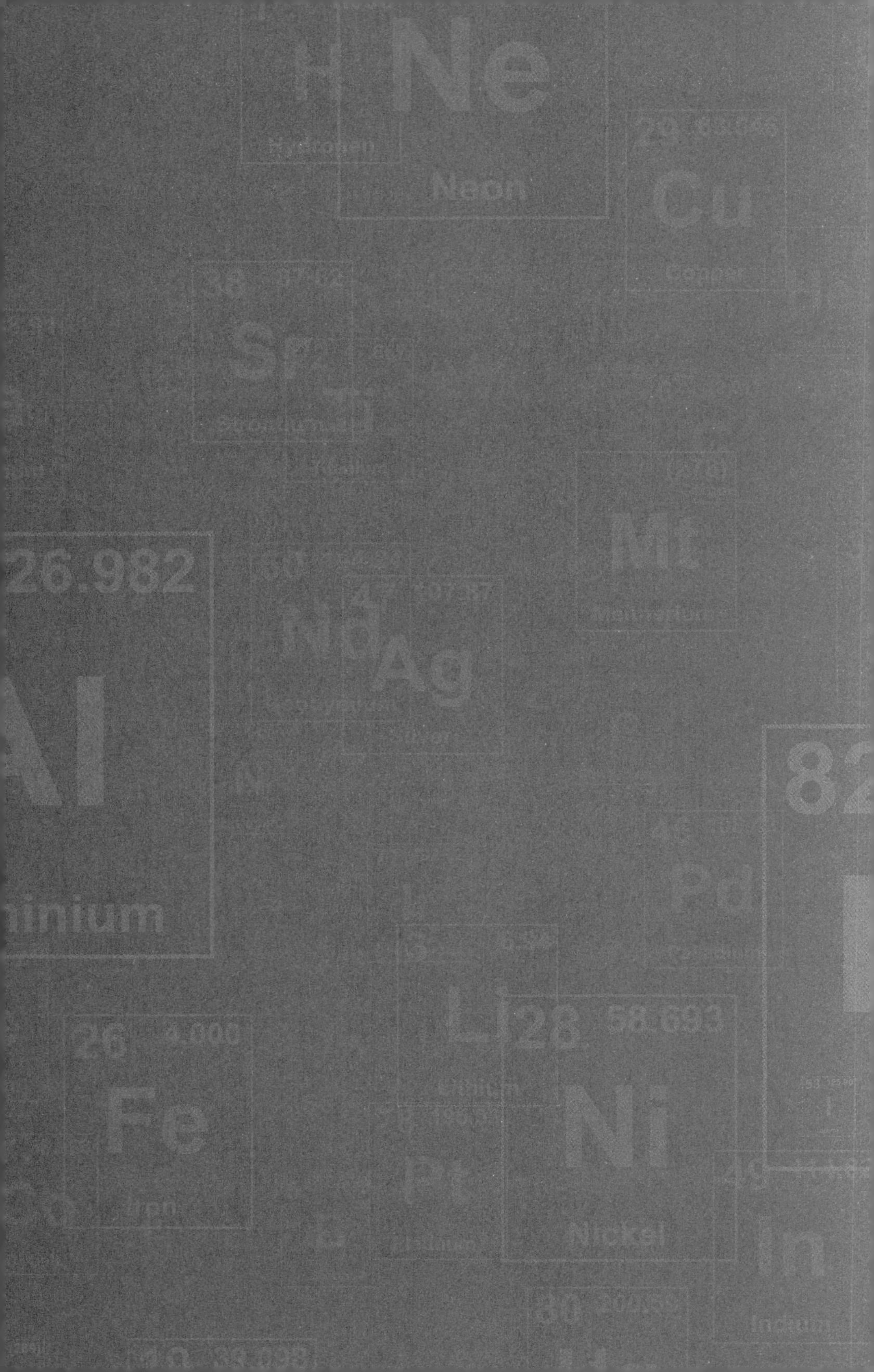

12

FROG MILKSHAKE

Before refrigeration, what were your options for a bucket of fresh milk on a hot day? You could either drink it immediately, or you could add time and turn it into cheese. But some farmers in Finland and Russia found another way to stop the milk from spoiling.

They would drop a frog into a bucket of milk, and chemicals from the frog skin would stop the milk-souring bacteria.

THE HISTORY OF COOLING

Of course, not everyone kept their milk fresh by chucking a few frogs into it. Some people kept it drinkable by using ice.

In days gone by, wealthy people in Ancient Rome or Ancient Greece would get a few tonnes of ice brought down from the mountains in late spring. They would then bury it in a hole in the ground and cover it with insulating material such as sawdust. *Voilà* – instant refrigeration that could last them until winter.

The first electrical refrigerator was invented around 1927. But the gases that did the actual cooling were a high-risk combination of poisonous, corrosive and/or inflammable.

This concept of ice as a Cold Store was re-invented in the USA in 1806 by Frederick Tudor, in Boston. He cut ice from frozen ponds with saws. He shipped 130 tonnes of ice to the tropical island of Martinique, and preserved it by using wood shavings, straw and blankets as insulation. His crystal-clear, bubble-free ice was even shipped to Australia.

Shortly afterwards, in the early 19th century, ice boxes became popular. An ice box was just an insulated box a bit bigger than a large microwave oven. It had a wooden frame, metal walls made from tin or zinc, and insulation such as sawdust, straw or cork. Every day, the ice man would come down the street with his horse-drawn cart loaded with ice and sell you a fresh chunk. That's how fancy city folk kept food cold.

The first electrical refrigerator was invented around 1927. But the gases that did the actual cooling were a high-risk combination of poisonous, corrosive and/or inflammable. Fridges became really popular in the 1930s only after Freon was invented. Freon was non-poisonous, non-corrosive and non-inflammable.

Unfortunately, its long-term effect was to punch a hole in our planet's protective Ozone Layer. It took until the 1970s to discover this. Since then, Freon and similar chemicals have been banned.

SAFETY FROG

Back to the small, poor, isolated rural villages in Russia and Finland who had no refrigeration of any kind. While it was freezing in winter, it got really hot in summer. Ice boxes just weren't an option for them.

Somehow, the villagers found they could stop the milk from going sour by putting a frog into the milk container. It wasn't just any frog – but the Russian Brown Frog, *Rana temporalia.*

What was going on? By itself, a bit of warmth won't turn milk sour. But milk is what the microbiologists call a great "culture medium". So when bacteria end up in milk, the warmth, fats, proteins and carbohydrates encourage them to grow like crazy. Lots of bacteria can make the milk sour.

Somehow, the farmers had discovered that plonking a Russian Brown Frog into a bucket of fresh milk would stop the bacteria from growing. The milk would get warmer – but it wouldn't spoil.

FROG SELF-DEFENCE

How do frogs stop bacteria? Well, every creature needs protection – from critters both smaller and bigger than itself.

Frogs are especially prone to bacterial attack, because their skin is moist – a perfect environment for bacteria to grow. All frogs release some kind of chemical

Plonking a Russian Brown Frog into a bucket of fresh milk would stop the bacteria from growing.

to keep bacteria at bay. Some of these chemicals are really good at fighting bacteria, while others can make you sick, or give you hallucinations.

In 2010, scientists from the United Arab Emirates (UAE) University discovered that the skin of some frogs would secrete or ooze out chemicals that could fight both bacteria and fungi. They showed that these chemicals were "peptides". Peptides are small proteins – just a bunch of amino acids strung together. The frogs studied by the UAE scientists were native to a few African countries.

Since then, we've discovered that a chemical found in the Mink Frog of North America can vanquish the dreaded "Iraqibacter". This is a drug-resistant bacterium that infected wounded American soldiers in Iraq.

As another example, secretions from the skin of the Foothill Four-Legged Frog seem to be able to fight antibiotic-resistant MRSA *Staphylococcus* skin infection.

RUSSIAN BROWN FROG

Peptides are small proteins – just a bunch of amino acids strung together.

The Old Wives' Tale – or the story passed down the generations by Elderly Female Domestic Engineers – about keeping milk fresh with frogs had been around for ages. But it took until 2012 before a team led by the organic chemist Dr Albert Lebedev, from Moscow State University, analysed chemicals on the skin of the Russian Brown Frog. They found some 76 different peptides.

About 27 of these peptides had significant antibacterial activity. They could kill gram-positive *Staphylococcus* as well as gram-negative *Salmonella* – both of which can be quite nasty to humans. Mind you, this antibacterial activity was seen only on the laboratory bench, not inside humans. But by a lovely coincidence, these peptides from the skin of the Russian Brown Frog also killed the bacteria that would land in milk and spoil it. (I have no idea how a farmer decided to try this out. Maybe it was serendipitous – a Russian Brown Frog jumped into a bucket of milk on a warm day and couldn't climb out, and someone noticed the milk didn't spoil?)

The scientists also found some 49 peptides related to a very interesting peptide called "bradykinin". Bradykinin is a peptide made of nine amino acids strung together. Bradykinin is also associated with the dreaded anaphylactic reaction where your airways close down, your face swells up so you have difficulty breathing, and your blood pressure drops. In some cases you can die from an anaphylactic reaction. Maybe these 49 peptides were to stop bigger creatures from eating the frog.

But getting back to dumping the Russian Brown Frog in your bucket of freshly produced warm milk: what about the problem of the frog urinating in the milk?

Well, we drink water from dams that fish live in, and fish wee, and that seems okay. So it stands to reason that a bit of frog wee is not going to make you croak . . .

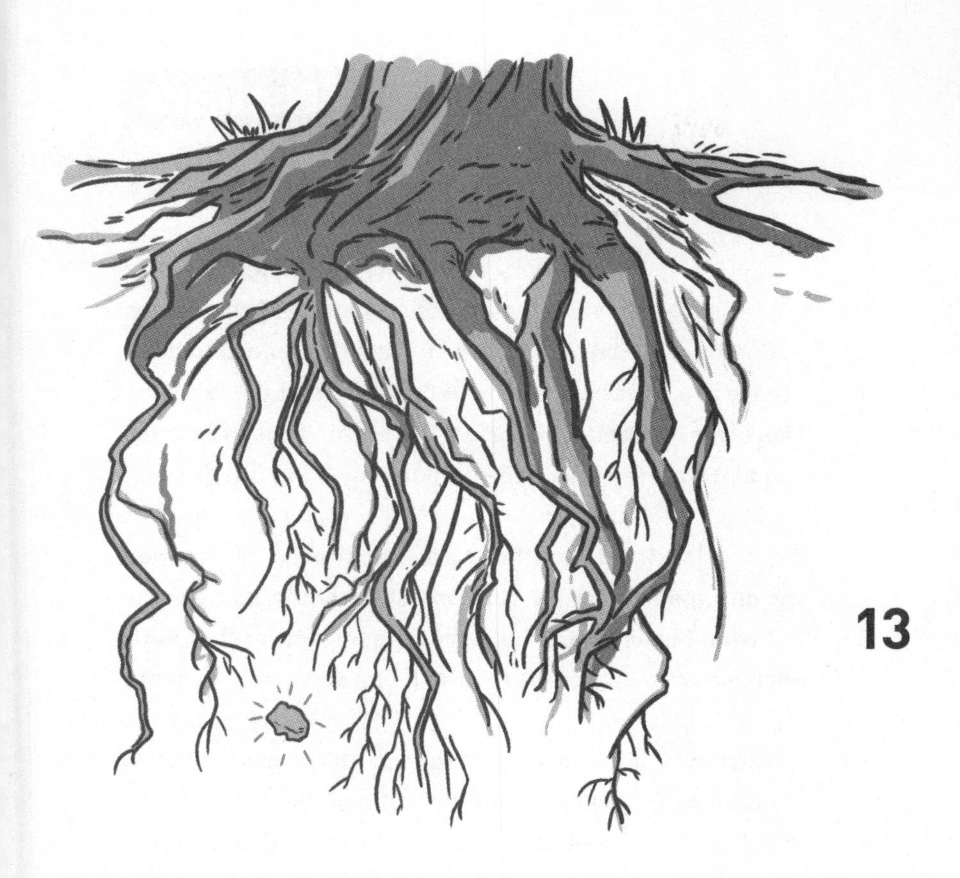

13

GOLD IN TREES

The seasons roll around, and when autumn comes many trees take on a gorgeous golden colour. But some Australian trees actually concentrate real gold (yep, the actual precious metal) into their leaves. How did it get there?

Gold itself started off at the Big Bang, and took a long journey via a molten planet Earth, giant asteroid impacts, bacteria, earthquakes, lots of geology and, yes, exploding stars. So there are a lot of steps to make from the Big Bang to the leaves of the eucalyptus tree – but here we go.

Disclaimer

I could be wrong in this story.

The story of gold drifts through fields as diverse as Cosmology, Microbiology, Seismology, Geology, Biology, and many more. With regard to gold, there are gaps in our knowledge in each of these fields. There are competing theories. Different areas of Science advance at different rates. We still don't have the full answer.

I've tried to gather all this scattered information into one coherent story. But almost certainly, some parts of what I have written will prove to be incorrect. (That's Science for you – five steps forward, one step back, repeat.)

Anyhows, this is my distillation of all those sources as of the middle of the second decade of the 21st century. I wonder what we'll reckon in the middle decade of the 21st century.

GOLD 101

We've been working gold for about 6000 years. The oldest goldwork dates to around 4000 BC – artefacts from the Balkans. We have simple Egyptian jewellery dating back to 3000 BC. Around 2000 BC, in Crete, Minoan goldsmiths crafted fine gold filigree work. The first golden coins come from Lydia in Asia Minor, from around 600 BC.

But there's not really much of the precious metal. All the gold we've mined over the last 5000 years would fit into a cube about 21 metres on each side.

Gold is definitely a "Precious Metal", with a price up around $60,000 per kilogram. Today, about half of the gold produced is used in jewellery, about 40 per cent for financial investments, and about 10 per cent for industrial uses.

Gold has many wonderful properties.

You don't need super-high temperatures to work it. It's easy to melt (at about 1064°C), but harder to boil (at about 2865°C).

It is very soft. In fact, if you beat a single gram of gold enough with a hammer, you can turn it into a transparent sheet with an area of one square metre. So it's very easy to turn into jewellery.

> Gold itself started off at the Big Bang, and took a long journey via a molten planet Earth, giant asteroid impacts, bacteria, earthquakes, lots of geology, and, yes, exploding stars.

It's a great reflector of heat, and so when it's coated as a thin film onto windows, it works effectively as an insulator. It's also a great conductor of electricity – only copper and silver are better. Ever notice the gold-plated contacts on the connectors of high-end electronics? That's because gold is both a great conductor – and very chemically inert. This means it won't corrode when used in electrical circuits – because very few chemicals will attack it.

Because it is so inert, back in the old days (several thousand years ago) people would find lumps of gold just lying around on the ground, still shiny and bright after thousands or millions of years out in the weather. (The other two elements found uncombined on the ground were silver and copper.)

But how did gold end up on the ground?

The Acid Test

Maybe you've heard someone talking about passing the "Acid Test". When a product easily satisfies its advertising claims, you can then say that the product passed the Acid Test.

The phrase comes from the fact that nitric acid would easily dissolve not only base metals such as iron, but even noble metals such as silver (which normally resist chemical attack).

However, gold (the royal metal) will not dissolve in nitric acid. So gold can pass the Acid Test.

Even so, *aqua regia* (which literally means "royal water") can dissolve gold. *Aqua regia* is a highly corrosive, fuming red or yellow liquid. It's a 3:1 mixture of hydrochloric and nitric acids. Gold also dissolves in alkaline solutions of cyanide and mercury.

COSMOLOGY DOESN'T GIVE YOU GOLD

In the Beginning was the Big Bang, some 13.8 billion years ago. At the start, the Universe was so hot that there was no "matter" as we know it today. The Universe was full of a super-hot plasma of fundamental particles and photons. (That "early soup" was very complicated.)

As the Universe expanded and cooled down, the first atoms coalesced into existence, around 377,000 years after the Big Bang. About 90 per cent of them were hydrogen atoms, about 10 per cent helium atoms – and a tiny fraction of other lighter elements.

At 377,000 years of age, there was no gold anywhere in the Universe.

A few hundred million years after the Big Bang, the first stars (which were much bigger than our Sun) were blazing away with their nuclear fires. These nuclear reactions involved lighter elements fusing together to make heavier elements – and the release of a huge amount of energy. Gradually, these early stars began making progressively heavier elements such as carbon, neon and oxygen – and then bouncing their way up through the periodic table to silicon and iron.

And still there was no gold in the Universe.

Supernova 101

A supernova is an exploding star that – briefly – is brighter than an entire galaxy.

There are two main types of supernova. One is the sudden reignition of nuclear fusion in a degenerate star (complicated). I'll talk about the other type: the gravitational collapse of a massive star (slightly less complicated).

First, what happens in gravitational collapse depends on the initial mass of the star – so let me pick a star around 100 times the mass of our Sun.

Second, that star is the size it is because of the balance of two mechanical forces. The forces that push outwards (radiation pressure from the nuclear burning process) are battling the forces that suck inwards (gravity, due to the enormous mass of the star). When the nuclear fuel runs out, the force of gravity takes over.

Within seconds the star collapses in on itself, bounces and expands outwards.

The core collapses inwards at about one-fifth of the speed of light (70,000 kilometres per second). The temperature and density increase enormously. New elements (including gold) are made. And then everything bounces outwards. The expanding radiation throws out much of the star's substance at speeds up to one-tenth of the speed of light (30,000 kilometres per hour).

ASTROPHYSICS GIVES YOU GOLD

Gold first appeared in the Universe when the first of these stars exploded in a supernova.

It turns out that, in general, gold can be "made" only in a truly cataclysmic event – as though the conditions at the centre of our Sun (15 million degrees Celsius) weren't hellish enough. Sure, our Sun burns the lighter elements to make heavier elements and energy. As a lucky side effect, this energy keeps conditions on Earth suitable for our kind of life. But when it does regular burning like this, our Sun is too "gutless" to make gold.

You usually need a supernova to make gold.

In the cataclysmic explosion of a supernova, atoms of gold are manufactured – and then hurled out into the Universe, along with the other debris from that explosion.

So exploding stars do two things for us.

First, they make most of the elements heavier than oxygen, and practically all the elements heavier than iron. These include, in addition to our favourite romantic and precious gold, more prosaic elements essential to the running of our bodies, such as sodium, potassium, calcium, zinc and iodine.

Second, they hurl these useful elements into space. Some of them end up in "Stellar Nurseries".

Gold from More than Exploding Stars?

It seems that gold may also be created when neutron stars collide. In addition, small amounts can be created as Red Giant stars get older.

But the overwhelming majority of gold is created when stars explode.

ASTRONOMY GIVES YOU SOLAR SYSTEMS (WITH GOLD)

Over time, vast "clouds" of gas and dust from these explosions began to coalesce thanks to the attractive force of gravity. Yes, there was gold in those clouds. The clouds formed what we call "Stellar Nurseries". A Stellar Nursery is a location where solar systems are born. (Yep, that is where baby suns come from.)

Inside these Stellar Nurseries, some locations are denser, while others are less dense (just random luck). A more dense location would have more gas and dust – and more gravity. As a classic example of a Feedback Loop, the greater gravity would attract more gas and dust, which would give it more gravity, which would attract more gas and dust, and so on and so on.

About 4.5 to 4.6 billion years ago, one rather special cloud evolved into a slowly spinning flattened disc. This would turn into our Solar System.

After about 10,000 years, this disc had gradually separated out into little lumps about 10 kilometres across, called planetismals. Over the next 100,000 to million years or so, these 10-kilometre-sized planetismals collided and stuck together. They formed smallish planets ranging from Moon-size to Mars-size – from one quarter the size of current Earth to about half its size. Then, over the next

10 to 100 million years, the larger planets formed. Specifically, it took about 30 million years for our home, the planet Earth, to form.

However, the majority of the gas and dust collected at the centre of this spinning disc, forming what became our Sun. Once enough matter had accumulated here at the centre, the temperature of this Sun-to-be got hot enough for nuclear fusion to happen. The nuclear fires ignited. The now-hot Sun at the centre of this spinning disc of planets and gas and dust just blew the gas and dust away – leaving behind our newborn Solar System.

Feedback Loops

Think of a microphone on a stage. The singer's voice goes into the microphone, then to the amplifier and, finally, out of the loudspeaker. Suppose the microphone gets too close to the loudspeaker. The sound leaves the loudspeaker and enters the microphone. At this stage the singer can stop singing.

The sound travels from the microphone to the amplifier and loudspeaker – and goes around and around. The sound gets louder and louder – with no further input from the singer. That's the distressingly loud Feedback Howl. It's called a Positive Feedback Loop.

"The rich get richer and the poor get poorer" is a classic example of a Feedback Loop.

But in the long term, in the Universe, it's so much fairer, and it all evens out. Suppose that a star gets bigger than its surrounding stars, thanks to a Positive Feedback Loop. Well, the bigger the star, the more chance it will explode.

MOLTEN EARLY EARTH = GOLD SINKS TO CORE

The early Earth, immediately after it had coalesced, had virtually no solid crust. Instead, the surface was mostly covered by a sea of molten rock and metals, such as iron. The Earth was molten all the way down to the centre.

Iron attracts the precious metals, such as gold and the platinum-group metals. Iron is denser than rock. Within some 5 million years, practically all the iron sank, dragging gold and other precious metals with it down into what is now the iron core of planet Earth.

There are enough precious metals inside Earth's core to cover the surface of our planet with a layer some 4 metres thick. Gold would make up half a metre of that 4 metres.

But, if the heavy metals sank down to the core, how come there's still some gold left near the surface, and how on Earth did it get into Australian eucalyptus trees?

Golden Fleece

I first heard of an actual, real, genuine Golden Fleece when I was in the highlands of New Guinea, in the early 1970s. I was in the old gold-dredging area of Bulolo, which had been very active in the 1930s. Even today, it's a remote and difficult area to get to. The first dredging machine was enormous, and weighed 1100 tonnes. Even so, the amount of gold at Bulolo was so great that it was financially viable to break down the dredge into its individual parts and fly the dredge parts into Bulolo by the tiny airplanes of the day.

The locals told me that in the old days, people would simply tie the fleece of a sheep to a tree and dump it into a river that was known for its rich gold panning. After some six months, the fleece would be heavy with gold. It would be burnt, leaving the gold behind.

In Greek Mythology, the Golden Fleece was the treasure that Jason and his Argonauts stole from King Aeetes. It was guarded by a fearsome dragon.

LATE HEAVY BOMBARDMENT DELIVERED GOLD

So how did the gold get back up to the surface of the Earth? Well, it hasn't. It's still down there.

The gold got delivered from space in the so-called Late Heavy Bombardment (which turned out to be the third set of "bombardments").

After the Earth first coalesced from the dust cloud, there was an initial period of large impacts by Moon-sized objects.

This was followed by a 500-million-year period of small impacts. Finally, the Late Heavy Bombardment occurred some 4.1 to 3.8 billion years ago. (This concept of the Late Heavy Bombardment arose after we analysed rocks that we got back from the Moon. Let me emphasise that this is still a new theory – but at this stage, it's fairly widely accepted.)

Gold first appeared in the Universe when the first stars exploded in a supernova.

Overall, we estimate that the Late Heavy Bombardment gave Earth some 20,000 craters bigger than 20 kilometres across, some 40 impact basins more than 1000 kilometres across, and several greater than 5000 kilometres across. Just think about that – 5000 kilometres is bigger than Australia.

The young Earth, then only half a billion years old, was "sprinkled" with heavy and precious metals delivered by some 200 billion billion tonnes of incoming asteroids in the Late Heavy Bombardment.

This added more heavy metals, including gold – but this time only to the crust, not deep below the surface.

Small Asteroid – Big Crater

Just as an aside, a relatively small asteroid can make a big crater. The Dinosaur Killer asteroid that slammed into the Yucatán Peninsula in Mexico some 65 million years ago was only about 10 kilometres across. But it made a crater about 250 kilometres in diameter, and 10 or so kilometres deep.

GOLD IN VEINS

Today, gold is mostly found in two situations.

There is primary gold in so-called "hydrothermal quartz veins", or just plain "gold veins". And then there is secondary "placer gold".

The traditional theory for the origin of secondary placer gold is that it came from the erosion of primary quartz gold veins. This erosion formed gold dust, gold flakes and small nuggets of gold – placer gold. This secondary placer gold is found in loose deposits of sand and gravel. Because the gold is much denser than the sand and gravel, it's easy to separate out – as in "panning for gold".

But how did the gold get into the primary hydrothermal quartz veins in the first place?

Current Gold Production

By the way, about one-third of our total gold production comes from direct mining of hydrothermal veins. About one-third comes as a welcome byproduct of producing copper, lead and zinc where the gold is an accidental contaminant.

THEORY 1: GOLD INTO VEINS VIA EARTHQUAKES

There are various theories about how gold "got" into the veins. But one recent and rather interesting theory is that earthquakes precipitated the gold into the veins.

This theory says that the hydrothermal veins formed (during periods of mountain building) over the last 3 billion years. Very large volumes of water (carrying dissolved gold) flowed inside seismically active faults.

An earthquake would make these faults suddenly expand. According to the simulations, a Magnitude 2 earthquake would increase the volume inside a fault by 130 times, while a Magnitude 6 earthquake would cause a much greater expansion of some 13,000 times.

This expansion would create a massive drop in pressure, and any liquid present in the vein would suddenly turn into a low-density vapour. Then quartz and various trace elements, including gold, would suddenly precipitate as a solid. Instantly, a quartz vein enriched with gold had been deposited – a mini goldmine-in-waiting.

In geologically active areas, repeated cycles of earthquakes over periods shorter than 100,000 years would increase the levels of gold in the veins by 1000 times – from two parts per billion to two parts per million. It's estimated that this process of earthquakes causing Flash Vaporisation could have created over 80 per cent of the world's economically viable gold vein deposits.

THEORY 2: PLACER GOLD VIA MICROBIOLOGY

There are a few standard theories about the origin of placer (or secondary) gold. Maybe the gold got there by erosion of gold-bearing rock, or maybe it got there by some kind of chemical activity.

But let's go out on a limb, and introduce a new (and not fully accepted) theory. Let's add "Life" to the story. Now we enter a new field of knowledge: Biogeochemistry.

Sometimes, when you look at placer gold with an electron microscope, it looks like piles of bacteria encased in gold. We have found what appear to be lacy patterns of bacteria in 2.8 billion-year-old South African gold, and in 220 million-year-old Chinese gold.

> There are enough precious metals inside Earth's core to cover the surface of our planet with a layer some 4 metres thick. Gold would make up half a metre of that 4 metres.

In 2006, we found a bacterium, *Ralstonia metallidurans*, which appears to "grow" gold nuggets from the surrounding dirt. These bacteria "pull" dissolved gold from the dirt around them towards themselves – and deposit it onto grains of gold that they live on. *Ralstonia metallidurans* are located inside a living biofilm on gold. (I wrote in great depth about biofilms in my 17th book, *Flying Lasers, Robofish and Cities of Slime*. At the time, 1997, the concept that bacteria could organise themselves into mini-cities was astonishing.) The *Ralstonia metallidurans* bacterium can survive in concentrations of gold that would kill other bacteria. Sometimes this bacterium accumulates the gold in specific areas internally (inside its cell), but at other times it is entirely externally covered by gold.

And in 2013, we discovered another bacterium that did the same trick – but by a different pathway. *Delftia acidovorans* manufactures a small protein that grabs any environmental gold and dumps it as a precipitate.

Maybe the various bacteria that "process" gold do so to get rid of this toxic substance. Most creatures can't survive high levels of gold. So if the bacteria works out how to live in association with gold, they can be fairly sure that they won't have to compete with other bacteria.

At this stage, we don't know how much of the gold that is dug up can be blamed on bacteria.

And now, finally we can get back to the gum trees that I mentioned in the very first paragraph. Welcome to the new field called Geobotany, or Paleobotany.

Metals In Plants

It's long been known, as a kind of Folk Wisdom, that certain plants above ground show proof of specific metals below ground.

Chinese prospectors of gold in Australia one and a half centuries ago apparently looked for certain plants they thought were associated with gold.

The West Australian geochemist Ian Crawford claims he found a big manganese deposit in Turkey by examining plants above ground. With regard to gold, he reckons that it's found not just in eucalypts, but in acacias and spinifex as well. The cost of drilling a single exploratory hole can easily reach $100,000. So he claims he uses plant inspection as a cheap "first pass" method to work out drilling strategies. Mr Crawford says he also uses plants to find other metals such as copper, zinc and nickel.

THERE'S GOLD IN THEM THAR TREES . . .

In Western Australia we have found eucalyptus trees that can extract dissolved gold from the water. The location (Freddo Gold Prospect, some 40 kilometres north of Kalgoorlie) is a hostile environment – semi-arid, with average maximum summer temperatures of 34°C. The rainfall is minimal (260–290 millimetres per year), and the trees lose a lot of water. (The evapotranspiration rates are greater than 2600 millimetres per year.) In other words, they drink (and evaporate) a lot, which means that they need deep roots to get a drink.

As the trees suck the water up, they also suck up dissolved gold. This gold comes from deposits some 40 metres below the surface.

We are very sure the gold has been absorbed from underground water and has not come from the wind simply blowing gold dust onto the tree from another site. The gold is present as tiny 8-micrometre-wide particles inside the cells of the tree. The highest concentrations are on the extremities of the tree, such as the distant leaves. Again, as with the bacteria, the gold is a toxin that the tree is probably trying to get rid of.

There's not a lot of gold in the trees. It would take the gold from over 500 trees to make just one gold ring. But it is very exciting that there is any gold in the trees at all.

And that's the story of how gold found its home among the gum trees. Hopefully, it has more than just a golden ring of truth to it.

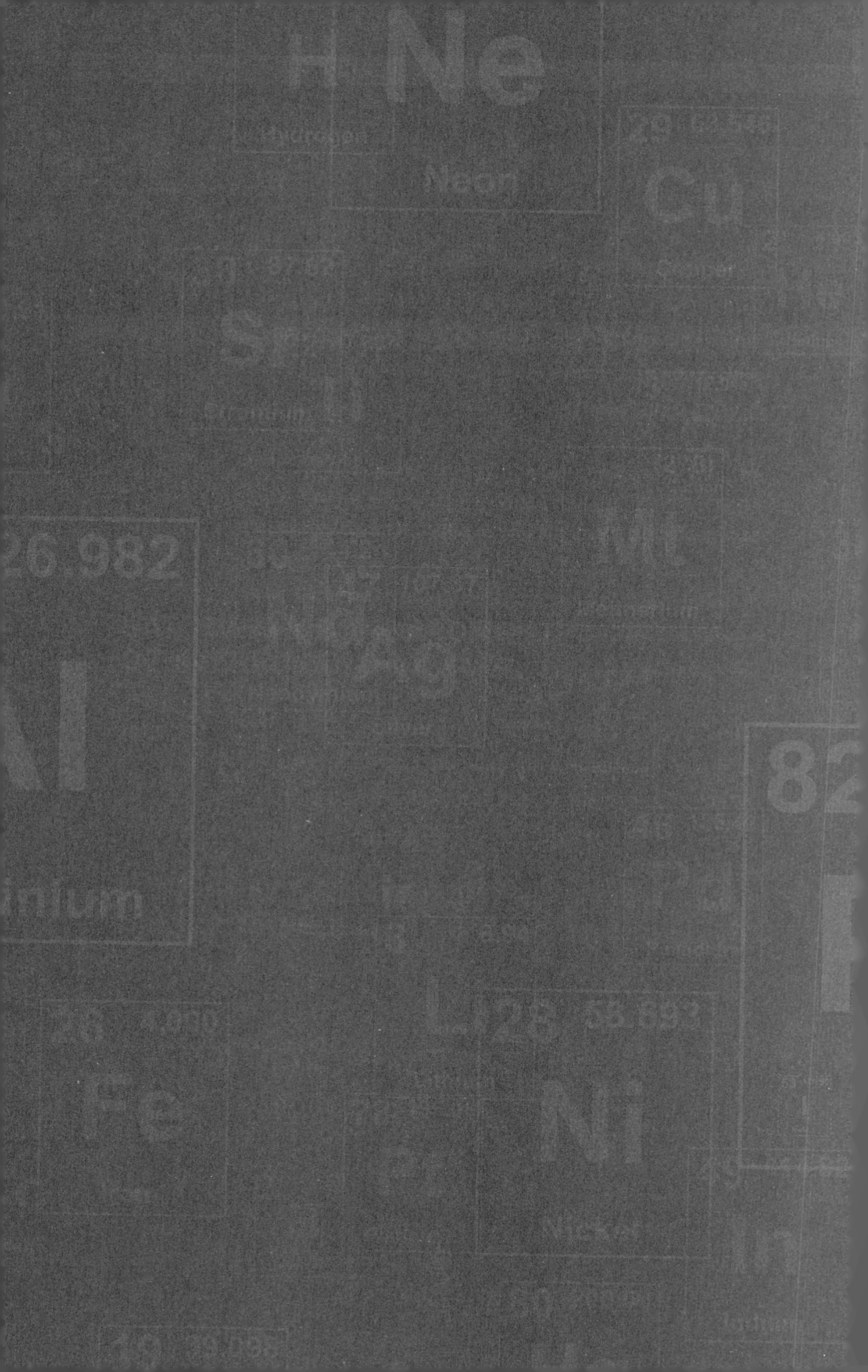

14

GREED IS NOT GOOD

In the movie *Wall Street*, the anti-hero, Gordon Gekko, preaches to an enraptured room, "Greed is good".

However, evidence tells us the opposite is true. Greed undermines moral behaviour. In the long term, that's bad – for the individuals it undermines, and for society as a whole.

This philosophy of "greed" being undesirable is not new. But it's easy to understand people wanting more. After all, the American singer Sophie Tucker supposedly once said, "I've been rich and I've been poor. Believe me, honey, rich is better." (This quote has also been attributed to Mae West, the comedian Joe E. Lewis, Frank Sinatra, Billy Connolly and many others. Perhaps this is a measure of how wise this quote is?)

What is Progress?

In 1937, the US President Franklin Delano Roosevelt gave his definition of "progress". He said that "progress" was making provision for the poor, not "[adding] more to the abundance of those who have much".

HISTORY OF GREED

A few thousand years ago, the Greek thinkers Plato and Aristotle postulated that greed, and desires for material wealth, would overpower one's original ethics. To summarise, they wrote that, "greed would be at the root of personal immorality".

Plato's and Aristotle's ancient wisdom has been corroborated by many studies, including a series carried out by Professor Paul Piff and colleagues at the University of California and University of Toronto.

Professor Piff et al carried out seven studies which looked at various aspects of the links between wealth and ethics.

Let me cut to the chase.

One of the co-authors, Professor Stéphane Côté, of the University of Toronto's Rotman School of Management, summarised their work with: "We found a trend that upper-class individuals – people who have the most money, the most income, the best education and the most prestigious job – have a tendency to engage in less ethical behaviour."

Wealthy Does Not Have To Be Greedy

There are many examples of highly ethical behaviour from upper-class folk.

Whistle-blowing is very difficult. Yet Cynthia Cooper (a former Vice President at Worldcom, which "fiddled" US$3.8 billion) pursued corporate whistle-blowing at great personal and professional cost. Further down the line, her actions enormously benefitted the greater good.

Furthermore, let's happily acknowledge the philanthropy of some fabulously wealthy people, such as Bill Gates, George Soros and Warren Buffett. They don't seem to be motivated by greed alone.

MODERN GREED

How did the researchers come to this conclusion?

In their first two studies, the experimenters left the laboratory to see what was happening on a daily basis at a busy four-way intersection with Stop signs on all approaches. They saw that drivers of more expensive cars were more likely to break the California Vehicle Code. These drivers broke the Code either by pushing their way through intersections, or by cutting off pedestrians at marked pedestrian crossings. About 50 per cent of Mercedes drivers did not give way to pedestrians, but practically all of the "lower-status" vehicles did yield. (Vehicles are, on average, a reasonable indicator of a person's wealth and social rank.)

The third study involved a series of eight different scenarios. It showed that upper-class individuals were more likely to act unethically or dishonestly in these eight situations.

High income inequality ripples through a society. It's strongly correlated with more crime, less happiness, less racial harmony, less civic and political participation, as well as poorer physical and mental health.

In the fourth study, upper-class participants "even took more candy from a jar that was ostensibly for children" when they were left alone with the jar for 30 seconds. What can you say?

In the fifth study, participants carried out staged negotiations. You can guess the trend – the more wealthy were more likely to tell lies.

In the sixth study, the upper class were again more likely to lie and cheat in order to increase their chances of winning a prize. And in the seventh study, people from upper-class backgrounds were more likely to endorse unethical behaviour at work.

What about a very basic crime – such as shoplifting? Other researchers have shown that people with incomes of greater than $70,000 per year are 30 per cent more likely to steal by shoplifting than their poorer cousins on $20,000 per year.

Yet another study has shown that poor people give almost 30 per cent more of their income to charity than do the middle class and the wealthy.

These kinds of results are consistent across multiple studies, by many different researchers, over the decades.

Inequality is Good/Bad for Business?

The overwhelming paradigm or wisdom in US politics over the past three decades has been simple – giving more aid to the poor and imposing higher taxes on the rich hurts economic growth. Research says this is incorrect.

On one hand, market economies need some inequality to work. But it seems that American economic inequality is so extreme that it's actually bad for business. As one of thousands of examples, consider talented children in poor families. They don't get the education they need to show their talent. It's unfair to the children, a waste of human resources and a cost to the country.

One of my heroes, Paul Krugman, says, ". . . there is solid evidence . . . from the International Monetary Fund, that high inequality is a drag on growth, and that redistribution can be good for the economy."

MASSIVE INEQUALITY – WEALTH 'N' HEALTH

What's going on? The rich are getting richer, and the poor are getting poorer. (I wrote about this in my 34th book, *Game of Knowns.*)

The USA is, from an economic point of view, the most stratified society in the Western world.

On one hand, some 14,000 American families (a microscopic top 0.01 per cent of the population) hold 22.2 per cent of the nation's wealth. That's nearly one quarter of all the wealth in the USA. Even during the recent Global Financial Crisis, wealth increased for the top 5 per cent of American families. Indeed, according to US Internal Revenue Service figures, "the richest 1 per cent has tripled their cut of America's income pie in one generation".

On the other hand, the bottom 90 per cent (more than 133 million families) have to share just 4 per cent of the USA's wealth – between all of them.

The middle class is missing out as well. In 2007, the "share of income going to the middle one-fifth of Americans shrank to its lowest level ever".

Poverty creates inequalities in health. In the USA, between 1983 and 1999, the life expectancy for women decreased in more than 900 counties – that's more than one quarter of all the counties in the USA. With regard to life expectancy, the USA is no longer on top – today, it's not even in the top 40 countries in the world.

Income inequality that is high ripples through a society. It's strongly correlated with more crime, less happiness, less racial harmony, less civic and political participation, as well as poorer physical and mental health.

Enough

Late renowned opthamologist Professor Fred Hollows taught me many things. One was the deceptively simple, "Most people go through life and simply don't know when they have got enough. That awareness comes only when they are on their death bed."

In their last days on Earth, very few people regret overtime shifts they didn't work.

THE POOR COPE

Poorer folk have fewer resources to draw upon, greater threats to their security, and greater uncertainty.

How do they get by?

A study by Professor Michael Kraus found that in person-to-person interactions, lower-class individuals were more likely to be able to accurately judge the emotions of another person. Taking it further and looking at photographs, poorer people were also more likely to detect which emotions another person was feeling. They did this just by looking at the muscles around the person's eyes.

> In person-to-person interactions, lower-class individuals were more likely to be able to accurately judge the emotions of another person.

Why?

Almost certainly it's because poor people have less control over their lives – so external forces often determine what happens to them. (They're kind of like a Paddle Pop stick in the gutter of life on a rainy day.)

As a result, they are more dependent upon other people in their lives. So it makes sense that poor people develop more skill in reading the emotions of the people around them.

As the penniless Blanche DuBois says in Tennessee Williams's play *A Streetcar Named Desire*, "I have always depended on the kindness of strangers."

THE RICH GET ENTITLED

So why does higher social class predict increased unethical behaviour?

One likely explanation is that the upper class have greater resources, greater freedom to act as they choose, and greater independence from others. This can give them "self-focused patterns of social cognition and behaviour". In plain English, they can get away with caring only about themselves – and sometimes they will.

And how do some of the wealthy respond to their privilege?

Paul Krugman won the Nobel Memorial Prize in Economics in 2008. He wrote, "self-pity among the privileged has become acceptable, even fashionable . . . a belligerent sense of entitlement has taken hold."

Today, governments are heavily influenced by the wealthy and the powerful. (Think of the benefits of "donating" to the re-election fund of a political party.) So the wealthy and powerful get to have laws and regulations that benefit them, to the detriment of the poor. But these influential people often lack empathy with others – and that goes double for the poor. But aren't governments in existence to govern for the good of the whole community?

Maybe it's not just that the wealthy have more dollars than sense, but that they also have more money than morals.

A Cure for Entitlement?

Paul Krugman's comment about the wealthy having a sense of entitlement is backed up by another study.

Professor Piff wrote, "higher social class is associated with increased entitlement and narcissism. Upper-class individuals reported greater psychological entitlement . . ." However, Professor Piff found a reason to be cheerful, because, ". . . inducing egalitarian values in upper-class participants decreased their narcissism to a level on a par with their lower-class peers".

15

GREEN TEA – LEADED OR UNLEADED?

That lovely warming beverage, tea, is one of the most popular liquids regularly consumed by humans. It gives you a "lift", without the edginess that can accompany a cup of coffee. There's one specific variety of tea called "Green Tea", about which many fantastical claims are made. But, as is the case with many health fads, not all Green Tea is what it's cracked up to be. In fact, in some cases, it can harm you.

First, there are indeed some powerful chemicals in Green Tea but it's not always true that "if a little is good, then more is better". Taking in too much of these chemicals can be toxic, especially to your liver.

Second, depending on where it's grown, Green Tea can be contaminated with toxins, such as lead.

So stick to drinking moderate amounts. It tastes nice, it might possibly have minor health-giving effects – and it's mostly harmless. Just avoid the concentrated extracts. And, by the way, I personally love the occasional cuppa Green Tea.

Caffeine in Tea

We all know that there's caffeine in coffee. An average cup has about 100 milligrams (but it can vary from 50 to 250 milligrams).

Black Tea has about 55 milligrams of caffeine, while Green Tea has much less – about 20 milligrams.

GREEN VERSUS BLACK (REGULAR) TEA

There are a few different types of tea around.

Black Tea is popular in Europe, North America and North Africa, and makes up about 80 per cent of all tea produced and consumed. (About 85 per cent of tea consumed in the USA is "iced" tea – a sugary imposter of what "real" tea is.)

Green Teas make up most of the rest of the tea that's enjoyed. It's widely consumed in China, Korea, Japan and Morocco.

Worldwide, only very small quantities of White Tea and Oolong Tea are consumed.

Interestingly, all these different types of tea come from the same plant, *Camellia sinensis*. This plant is grown mostly at altitudes of one to two kilometres, and mostly between the Tropic of Cancer and the Tropic of Capricorn.

There are two main variables as to how different teas arise. First, Green and Black Tea tend to come from different varieties of *Camellia sinensis.*

The second variable is how the tea leaves are processed. For Black Tea the leaves have been oxidised – for Green Tea they have not.

Catechins

Catechins are chemicals. They belong to a group of chemicals called "polyphenols". They're the chemicals in Green Tea that are most often claimed to have health benefits.

In test tube studies (but not as convincingly in human population studies), catechins have antioxidant, anticarcinogenic and antimicrobial properties.

Unfortunately, we have been told the simple mantra that "oxidants are bad, antioxidants are good". The reality is much more complicated. For example, some oxidants attack cancerous cells.

The same goes for microbes – they are not all bad. Indeed, 90 per cent of the cells in your body do *not* have your DNA – they are microbes. You don't want to kill them, as they help keep you healthy. (This is discussed in my 31st book, *Brain Food.*)

CHEMISTRY OF TEAS

The dried leaves of both Black and Green Teas contain roughly the same levels of proteins and amino acids (about 19 per cent by weight), fibre and other carbohydrates (33 per cent), fats (7 per cent), minerals (5 per cent) and pigments (2 per cent).

Here's the difference. The Black Teas have gone through an oxidising process for up to four hours. (Oolong Teas get only two hours of oxidising.) By weight, Black Teas contain about 25 per cent "oxidised phenolic compounds" such as theaflavins and thearubigins, but only 5 per cent "non-oxidised phenolic compounds". These make the liquid tea darker in colour.

However, the Green Tea leaves have not gone through any oxidising process. (Neither have the White Tea leaves.) As a result, their leaves contain about 30 per cent non-oxidised phenolic compounds, such as catechins. Green Tea is lighter in colour than Black Tea.

BENEFITS OF GREEN TEA?

> Green and Black Tea tend to come from different varieties of *Camellia sinensis.*

It's mostly the non-oxidised phenolic compounds that are purported to be the basis of all those amazing health benefits of Green Tea. They make Green Tea an easy-to-market treatment for headaches, various body aches and pains, and depression. It's also claimed that Green Tea will protect you against cancers of the stomach, colon, breast, ovary, bladder and prostate. And just to round things off, there are also claims that Green Tea protects against osteoporosis and dental caries.

One fact to consider is that many of these supposedly beneficial chemicals in Green Tea are also found in fruit and vegetables, and in cocoa and wine. (Clearly, there should be more health-food stores specialising in chocolate and wine?)

The second thing is that the overwhelming majority of studies are not conducted on humans. Instead, they're mostly either animal studies or cell-line studies on the laboratory bench. This makes the results less than conclusive.

HUMAN POPULATION STUDIES

There are some human studies dealing with the health benefits of Green Tea. But, from a statistical point of view, they are mostly not very rigorous. They are predominantly low-grade observational studies – not proper double-blind controlled studies. So they tend not to account for factors that could alter the results, such as family medical history, family wealth, diet and so on.

Some studies show that consumption of Green Tea is associated with a reduction in coronary heart disease and ischaemic stroke. A few other studies show improvement in diabetes.

However, the vast majority of the human studies can be interpreted as showing that people who have healthy lifestyle habits tend to drink Green Tea. In other words, drinking Green Tea is more a result of the healthy lifestyle than the cause of the health benefits. This is a classic case of confusing Correlation with Causation.

Furthermore, the human studies have conflicting results. Some studies identify various health benefits, but other studies can't back them up.

The summary is simple – the health evidence for Green Tea is weak. On the other hand, Green Tea does taste nice. This is as good a reason as any to drink it.

LEAD IN GREEN TEA

Tea is usually grown in highly acidic soils. This acidity makes any lead present in the soil more bioavailable to be picked up and absorbed by the roots of the tea plant. Tea plants also have a relatively large leaf area.

This makes the plant more able to pick up lead from the atmosphere.

Currently, about 80 per cent of Green Tea is produced in China. A study in 2006 measured lead levels in Chinese Green Teas. It found that some of them were 50 times the maximum permitted level. Overall, there was an increasing trend in lead levels found in tea over the years 1989 to 2000. Lead contamination is due to China's massive industrialisation during this period. Unleaded petrol was introduced into China as recently as 1993, and the sale of leaded petrol was stopped only in mid-2000. Since then, it seems there has been a downward shift in lead levels in Green Tea.

The lead levels have been found to be higher if the tea plants are near a highway, or if the surface layer of the soil is contaminated. The longer before the tea leaves are picked, the higher the lead levels. To give you an idea of the numbers, the lead levels are also two to three times higher in older leaves than younger ones. (By the way, lead contamination is not really a problem with most Japanese teas.)

On the other hand, there is some good news. The lead in the Green Teas that ConsumerLab.com, an independent tester of health products of all kinds, tested tended to stay on the tea leaves. It didn't go into the Green Tea liquid – keeping tea connoisseurs relatively safe.

Lead – Universal Toxin

Lead is a powerful physiological and neurological toxin that affects virtually every organ and system in the human body – including the reproductive system and the kidneys. The most sensitive is the Central Nervous System – unfortunately, especially in children.

Lead reduces intellectual performance and cognitive development in children.

NATURAL TOXINS IN GREEN TEA

Some of the chemicals naturally present in Green Tea are quite powerful. Drinking low-to-moderate amounts of Green Tea is safe. But drinking huge amounts of Green Tea, or taking some kind of concentrated Green Tea extract, is not.

Green Tea can reduce iron absorption. It contains Vitamin K, which can make "blood-thinning" drugs such as warfarin less effective. It can also interact with benzodiazepines and diuretics.

There are cases on record of Green Tea extracts causing kidney damage, as well as liver damage and visible jaundice. Many of these cases required hospital admission.

> Tea is usually grown in highly acidic soils. This acidity makes any lead present in the soil more bioavailable to be picked up and absorbed by the roots of the tea plant.

In some cases, a Health Spa has put people onto a fast consisting of drinking large quantities of Green Tea. Unfortunately, fasting increases the bioavailability of some of the chemicals in Green Tea, such as EGCG – epigallocatechin-3-gallate. (EGCG is an antioxidant, and has some anti-cancer properties. But oxidants are not always bad, and antioxidants are not always good.) In high levels, EGCG can cause health problems.

DIFFERENT VERSIONS OF GREEN TEA

Americans consume about 10 billion servings of Green Tea each year in various forms. In 2013, when ConsumerLab.com tested Green Teas, two of the eight supplements failed the testing procedures. One supplement contained 136 milligrams of caffeine per capsule. The recommended daily dose of three capsules would be the equivalent of four cups of regular coffee. Another supplement contained only 83 per cent of the claimed amount of EGCG.

The pre-prepared bottled Green Tea beverages were mostly loaded with sugar and extra caffeine. Some of them didn't carry the advertised quantities of "good" Green Tea chemicals as claimed on the label. In other words, some were mostly sugared caffeine water with a misleading label.

The chemical analyses of the loose leaf and tea bag versions of Green Tea were better. They did tend to carry more of the expected Green Tea chemicals. The surprise was that some of the tea bags delivered 2.5 micrograms of lead per serving.

Green Tea was apparently first brewed way back some 28 centuries before Christ, in the reign of the Chinese Emperor Chen Nung – so it has a long history. Maybe the Take-Home Message is to follow traditional tea procedures. Sip it demurely, rather than guzzle down gallons.

Anecdotal "Testimonial"

Here is an email I received, giving a typical account of an overdose of Green Tea. As expected, the delivery method was capsules, not liquid tea. It's hard to overdose with tea, but regrettably easy with capsules. Note that the writer was taking the *recommended* daily dosage of supplements. In this anecdotal case, the organ most damaged was the liver. This fits in with the known liver risks (in the medical literature) of excessive consumption of Green Tea extracts.

". . . I was taking an Australian branded Green Tea supplement purchased from a reputable health food store, not for weight loss, but for the health benefits that I'd read so much about. After taking the recommended daily dosage of supplements for 5 months, I contracted 'toxic hepatitis'. Fortunately, I didn't need a transplant, but my liver was damaged, I was jaundiced, I had to have a biopsy, I lost a third of my hair, and had to take corticosteroids for a few months before my liver enzyme tests returned to normal (a normal range being 20 to 30, but at the height of my illness, my levels reached 3300)."

Luckily, the writer recovered in full. Green Tea is mostly harmless, if you drink the tea itself. Avoid the concentrated extracts.

16

LACTOSE INTOLERANCE

Food can bring people together. But some foods can be problematic for dinner guests.

A good example is milk, and the condition known as "Lactose Intolerance". But first, let me be very specific about Lactose Intolerance. This is a normal state of affairs for two-thirds of adults. (Maybe you personally know only a few people who are Lactose Intolerant? The explanation is related to Geography. Read on.)

How did we end up with the situation where about two-thirds of the adults on our planet can't comfortably drink milk? The answer is a combination of Cattle Herding and Evolution.

Milk is Bad?

Some people claim that "milk is wrong" because no other mammal drinks milk as an adult.

Well, no other mammal has developed poetry, income tax or string quartets. Does that make them wrong?

LACTOSE AND LACTASE 101

First, what is the different between "lactose" and "lactase"? Lact*o*se is a carbohydrate (or sugar), and lact*a*se is a protein (or enzyme) that splits lactose. (In Chemistry, a chemical whose name ends in "ase" is an enzyme. Enzymes are proteins that speed up other chemical reactions.)

The Organic Chemists call lactose a "disaccharide", which literally means "two simple sugars". Lactose is the common sugar in cow, sheep, buffalo, goat and human breast milk. It is actually made of two simpler sugars joined together. These simple sugars are glucose and galactose. They have the same number of atoms of carbon, hydrogen and oxygen – but the atoms are arranged slightly differently.

Lactose is physically too large to cross the gut wall and enter the bloodstream. So all young suckling mammals manufacture a special enzyme to split the lactose into the two simpler sugars – glucose and galactose. These sugars are small enough to cross the gut wall and enter the bloodstream.

This enzyme, called "lactase", is found in the cells that make up the lining of your jejunum and duodenum – parts of your small intestine.

Thanks to lactase, baby mammals can get all the nutrition they need from their mothers' milk. Besides carbohydrate, milk includes water, fat, protein, minerals, immune system chemicals, etc. Breast milk is a so-called "complete food". But growing mammals generally lose this enzyme after they stop suckling.

Today, two-thirds of human infants stop making this enzyme somewhere around 3 to 5 years of age. As far as we currently know, until several thousand years ago, no adult mammals could manufacture this enzyme.

Milk Allergy

There is a medical condition called "Milk Allergy".

This has nothing to do with lactose – which is a carbohydrate. Instead, Milk Allergy is a specific reaction to certain proteins in the milk.

BUT 12,000 YEARS AGO . . .

Humans like us, *Homo sapiens sapiens*, have been nomadic or semi-nomadic for some 200,000 years.

But about 12,000 years ago, somewhere near modern-day Turkey/Iran/Iraq, we began the long and complex process of the Neolithic Transition.

Around that time, people acquired the social behaviours needed to leave a nomadic lifestyle and adapt to a sedentary way of life. They developed new economic strategies, such as animal and plant domestication. They realised new technologies, such as pottery and polished stone tools. Around 11,000 years ago, somewhere between the Zagros Mountains in present-day Iran and Iraq and the Taurus Mountains in present-day Turkey, they started domesticating goats and sheep. Cattle and pig domestication began about 500 years later.

The practice of domesticating animals reached present-day Romania and Hungary about 7500 years ago. It took a little longer to arrive in Britain, crossing the English Channel 6000 years ago.

ANIMAL DOMESTICATION

When you domesticate an animal, you can collect either "primary" or "secondary" products.

> Lactose is a carbohydrate (or sugar), and lactase is a protein (or enzyme) that splits lactose.

Primary products include meat, bone, hide, horn, and so on. However, you can get these primary products only by killing the animal. Secondary products – such as wool, the animal's dung, its labour and, yes its milk – are different. You can continue harvesting secondary products over the life of the animal.

At this stage I imagine you're asking for more stats regarding the milk production of a prehistoric or Neolithic cow. Well, it takes about eight months for a calf to be weaned. Suppose that the mother has access to good food, and that you stop the baby calf from drinking all the mother's milk. After you subtract the milk needed to raise the calf to independence, you're still left with some 200 kilograms of milk over that period. This milk gives you about the same "energy" you would get by eating the meat of the whole cow.

You can eat a cow only once, but with clever husbandry, you can milk the cow for years. Cows are the goose that lays the golden egg.

MILK – FOR AND AGAINST

There are many advantages to consuming the milk of a cow.

First, it is a renewable source of food energy – and for our ancestors, times could be bad. After all, cows eat grass, which is readily available, and which we humans can't digest anyway. Cows and humans don't compete for the same foods.

Second, milk is mostly water. This was very handy in a drought.

Third, milk is not usually contaminated with nasty bacteria. Yes, the udders are near the cow's anus, but you can be careful.

And fourth, if you live near the Arctic, it contains some useful Vitamin D as well as lots of calcium.

But there was a big problem for our ancestors.

Back then, only suckling babies and infants could digest milk.

They were the only humans who naturally had the lactase enzyme in their small intestine. Remember – lactase is essential to break down the indigestible two-sugar lactose into two separate digestible simple sugars.

If adult humans tried to drink milk, the big disaccharide molecule of lactose would not get digested or broken down in the small intestine, but instead, would continue on into the large intestine. There, various microrganisms (including bacteria) would try to break it down – with varying degrees of success. Symptoms would include stomach discomfort, excessive farting, cramps and bloating. Indeed, lactose molecules in the colon can have an osmotic effect and draw in water from the bloodstream, leading to soft stools or even diarrhoea.

Initially, adults didn't benefit from drinking cow's milk. But at the very least, the cows' milk could keep the suckling infants alive during bad times.

Not Common Lactose Intolerance

There are two medical conditions slightly related to Lactose (the sugar) Intolerance – which is actually Lactase (the enzyme) Deficiency. Again, let me emphasise that Lactose *In*tolerance is more common (two-thirds of humans) than Lactose Tolerance (one-third of humans).

One condition is called "Congenital Lactase Deficiency". (Remember that the protein enzyme lactase breaks down the carbohydrate called lactose.) It's a very rare genetic disease in which the lactase enzyme is simply not present – even immediately after birth. The good news is that if these newborn babies can get a milk formula that does not contain lactose they grow up just fine.

A bit more common is the medical condition called "Secondary, Acquired, Or Transient Lactase Deficiency". It happens occasionally as a result of some kind of injury to the small intestine, such as from intestinal parasites, chemotherapy or acute gastroenteritis. (Remember that the lactase enzyme is present in the wall of the small intestine.) It usually resolves by itself.

EVOLUTION TO THE RESCUE

Of course, in pre-Lactose Tolerant peoples, there would be a lot of natural human variability in how well they could digest milk. Some adults would cope better than others.

How Lactose Intolerant people coped would depend on how much lactose they consumed in one sitting, how efficient their natural gut bacteria were at breaking down lactose, and so on.

For example, suppose that at the same time that they drank milk, they also ate other foods (especially fatty or high-energy ones). The stomach passes its contents on to the small intestine at about 8 to 12 kilojoules per minute. Delayed by the other foods, the milk would pass into and through the small intestine at a slower rate, which would give the bacteria more time to process it.

> Lactase is essential to break down the indigestible two-sugar lactose into two separate digestible simple sugars.

Human ingenuity also played a part. We've been making cheese for at least 8000 years – and probably 10,000 years. We've been using inflated sections of animal gut as storage vessels for millennia. A chemical called "rennet" is found in the stomach of animals. By a wonderful coincidence, rennet turns milk into curds (solid) and whey (liquid). Process the curds and you have cheese. Many cultures have legends about how an ancestor used an inflated stomach to store milk – and accidentally invented cheese.

Cheese has two advantages over milk for the Lactose Intolerant. First, it's lower in lactose. Second, its higher fat content slows its transit time through the gut, and makes it easier to digest. Fermented cheeses such as cheddar and feta are quite low in lactose, while aged hard cheeses such as Parmesan have even less.

But what our ancestors really needed was some good old Evolution. This allowed them to manufacture lactase as they used to be able to, back when they were breast-feeding infants.

And so, beginning in both Europe and Africa, our ancestors evolved the ability to make the enzyme lactase.

The Evolution of Lactose Tolerance (It's a Bit Complicated)

Let's talk genetics – specifically, about two genes located on the long arm of Chromosome 2.

One gene (LCT) holds the instructions for making lactase. The other gene (MCM6) can turn LCT on and off. It's a control gene.

A mutation on this control gene (MCM6) arose some 7000 years ago. It happened in the geographical region around modern-day Hungary. It kept the lactase gene turned permanently on, so these cattle herders – as adults – could now drink the milk from their cows.

Quite separately, on at least three occasions in at least three different locations in Africa, slightly different mutations occurred. These separate (and different) mutations happened in distinct populations of cattle-raising peoples – some 3000 to 7000 years ago. That 4000-year period is very short – just the blink of an evolutionary eye.

Just like the European mutation, these African mutations had exactly the same result: keeping the lactase-making gene turned on.

LACTOSE TOLERANT ADULTS MULTIPLY

The short-term result of Lactose Tolerance in adults was an enormous survival advantage. They got more "nutrition" from the same number of cows. As a last resort, they could eat the cows. They could survive longer in bad times – longer than their Lactose Intolerant cousins.

The long-term result was that the Lactose Tolerant cattle-herders could have about 10 times as many fertile descendants than the people who had lost the ability to make lactase after infancy.

This mutation to be able to digest milk as an adult was so useful that it spread very rapidly.

In terms of Geography, the closer you are to where the mutations arose, the greater the percentage of people who are Lactose Tolerant. Some 99 per cent of Swedish and Dutch people are Lactose Tolerant – but as you move further away, the percentage drops. It drops down to 50 per cent in Spanish, French and Arab populations, and to only a few per cent in China.

Evolutionary Biology

The cattle herders in Europe and Africa each had quite different gene mutations. They each took completely independent pathways. But they all converged on the same end-point of Lactose Tolerance.

Geneticists see Lactose Tolerance as an extremely elegant example of "Convergent Evolution". This occurs when Nature "finds" several different solutions for the same problem.

Now, you'd normally think that evolution happens because of changes in the natural environment. For the European and African pastoralists, part of their "natural" environment was their cattle. These cattle then drove the genetic adaptation in people. This is the so-called "Culture-Historical Hypothesis".

We have found traces of milk proteins in ceramic vessels in present-day Romania and Hungary. These date back 7500 to 8000 years. We've found in Poland some three dozen perforated pottery vessels, also about 8000 years old. They were used to strain milk as part of the cheese-making process.

Lactose Intolerance is not black and white. In some cases, even though your DNA does not make lactase, the bacteria in your gut does. For example, Lactose Intolerant Somali people living in Ethiopia today can drink 500 millilitres of milk each day without any obvious discomfort. Today, some 40 million Americans are affected by Lactose Intolerance to some degree.

We can all break down starch in carbohydrates, but not all of us can split alcohol. Now that we have started tracking the evolution of lactase, maybe we can find the origins of amylase (an enzyme that breaks down starch) or even alcohol dehydrogenase, which lets us break down alcohol.

And then we'll be able to understand why the Blues singer, John Lee Hooker, in "It Serves Me Right to Suffer" could tolerate the doctor putting him on milk, cream and alcohol . . .

17

MEAT ALLERGY

There seem to be many immediate health benefits to a well-planned vegetarian diet – and you can live longer as well. But in most parts of Western Society, vegetarians are in the minority – say, between 2 and 10 per cent of any given population. But this could change thanks to the rapidly increasing spread of an Australian tick. For some people, one bite from this tick makes them allergic to meat – for the rest of their lives.

I'm sorry – enough with the scare tactics. (I shouldn't write like the tabloid gutter press does. I won't be selective with the facts. Sorry.)

These once-bitten people can still eat some meats, such as fowl and fish. It's just non-primate mammals – such as beef – they can't eat.

TICK OF DOOM

The offending tick is the Australian Paralysis Tick – *Ixodes holocyclus*. It lives along the east coast of Australia, from Lakes Entrance in Victoria to well past Cooktown in Far North Queensland – some 3000 kilometres of ocean frontage. The adult ticks are about 3 to 10 millimetres across, and shaped like a human fingernail. They attach themselves to plants, and jump onto a passing bandicoot or human. They'll crawl up inside your clothing and get to your blood by biting through your skin, usually on the head and neck.

> The best way to remove the tick is to first stun or freeze it.

There are three major possible outcomes when they bite you. These are (1) hardly anything, (2) a mild allergic reaction, or (3) a severe allergic reaction.

In most cases, you will get a little local itching and swelling. It's unpleasant, but this is not an allergic reaction.

Sometimes there will be inflammation (redness) at the site of the tick bite combined with a large swelling of the flesh, and both the swelling and inflammation will last for several days. In this case, you've had a mild allergic reaction to the tick bite.

But every now and then the major life-threatening allergic reaction known as "Anaphylaxis" occurs. Anaphylaxis can cause a rash all over your body and swelling of the tongue and throat so serious that you have difficulty in breathing. It can also set off vomiting and diarrhoea, and a serious fall in blood pressure. In short, you are at major risk of dying. Anaphylaxis is a medical emergency.

The Best Way to Remove a Tick

Do not try to kill the tick with insecticide, oil, turpentine, kerosene, etc. These will usually irritate the tick, which may respond by injecting more allergen-containing saliva into you.

The best way to remove the tick is to first stun or freeze it. So the advice for people who live in tick-prone areas is to use an ether-containing engine-start spray pack, such as Aerostart, or a high-alcohol substance. This stops the tick from injecting its saliva. Then you can, at a more leisurely pace, scrape the tick out.

THE MYSTERY

Back in 2007, Associate Professor Sheryl van Nunen, an Immunologist at Sydney's Royal North Shore Hospital, wrote about her remarkable observations. Her paper discussed a small cluster of 25 patients who had been bitten by the Australian Paralysis Tick.

After several months of eating meat with no problems at all, out of the blue, they each had an allergic reaction after eating meat. The main meats that caused problems were beef (not because it's especially "dangerous", but simply because it's so prevalent in our society), lamb and goat. However, venison, buffalo and kangaroo set off an even stronger reaction. Interestingly, the reaction was stronger if the meat was rarer.

Seventeen of the 25 patients had one or more of the symptoms of rash, fainting, swelling of the tongue, constriction of the throat, and difficulty in breathing combined with an audible wheeze when they tried to breathe. These symptoms can be extremely distressing, and sometimes life-threatening.

Surprisingly, none of the patients had reported a major allergic reaction to the original tick bite. Although, all except one had reported a stronger than normal reaction to the initial tick bite. At the time of the tick bite, they had each developed a welt bigger than 5 centimetres across.

THE ANSWER

Finally, after a lot of work, the immunologists seem to have worked out the cause of the meat allergy. (By the way, immunologists tend to be really clever doctors, simply because Immunology is so fiendishly complex and difficult to unravel.)

It starts with a not-very-sweet sugar called "galactose". We humans can eat it without any problems.

When you combine galactose with the sugar called glucose, you get lactose. That's the sugar in breast milk. (See "Lactose Intolerance" on page 151.)

The trouble begins when you combine two galactose molecules together in a rather special way to make a bigger sugar called "galactose-a-1,3-galactose" – commonly called Alpha-Gal. All mammals carry Alpha-Gal – except for humans and the higher primates. In fact, it turns out that we can be allergic to it.

Allergic to Anti-Cancer Drug?

An early hint on the role of Alpha-Gal came from the USA. An anti-cancer drug, cetuximab, was being used in the USA. Researchers began to report an unusually high rate of allergic reactions to this drug – on its first use.

It turned out that cetuximab contained Alpha-Gal, due to how it was manufactured.

So, almost certainly Alpha-Gal is the cause of the mammal meat allergy.

But how does it get into you?

TICK + MAMMAL = ALPHA-GAL IN YOU

So here's the scenario.

A bandicoot (or another cute animal) is enjoying life somewhere on the east coast of Australia. The Australian Paralysis Tick bites it to get a meal of blood. Some of the Alpha-Gal from the bandicoot gets transferred into the gut of the tick.

After a while, the tick feels hungry again, and goes on to bite a human. Some of that Alpha-Gal from the bandicoot has just entered the bloodstream of a human.

Many, many Australians get bitten by ticks, but very few then go on to get an allergy to mammalian meat.

Something happens to that human who becomes allergic, (we still don't know what) and their Immune System slowly creaks into action. The specific immune system chemicals that are involved are Immunoglobulin E antibodies. After a delay (somewhere between one and six months after the initial tick bite), the human has yet another regular meal involving meat – but on this specific occasion, they get an allergic reaction. In some cases, they can die from a full-blown Anaphylactic Reaction – unless they can get medical treatment in time.

All mammals carry Alpha-Gal – except for humans and the higher primates. In fact, it turns out that we can be allergic to it.

This allergic meat-reaction will be set off by pork, beef, lamb or even whale meat. But fish or chicken are safe to eat because they are not mammals. The reaction can also be set off by some marshmallows, if they contain beef gelatin. (Yep, the Universe is a dangerous place …)

IT'S SPREADING . . .

Today, Associate Professor van Nunen is seeing about two cases each week of tick-caused meat allergy, and has over 500 patients with this condition on her books. In the USA and Europe, there are thousands of people who have become allergic to mammal meat as a result of a tick bite. The tick can be the Lone Star Tick in the USA, or the Castor Bean Tick in Europe.

The reason that it's increasing might be a change in tick or mammal behaviour, or simply better reporting and understanding, or something else. We're not sure.

So in more ways than one, these people could literally (and I do mean literally) die for a steak.

Alpha-Gal Coincidence?

Alpha is the *first* letter of the Greek alphabet.

Alpha-Gal got its name from the location on the galactose molecule where the two galactose molecules linked to each other. But Alpha-Gal allergies are the *first* known allergies to be linked to a sugar (carbohydrate), rather than a protein. They are also the *first* food allergies to have a delayed Anaphylactic Reaction.

Is this all just a simple coincidence?

Consider this. There are 24 cans in a slab of beer, and 24 hours in a day. Is that also just a coincidence?

18

NAZIS STOLE SPACE BUDDHA

Good titles grab your interest. Book publishers know that certain buzzwords will make people even more inclined to buy your book. According to Alan Coren back in 1975, *Nazi Golfing for Cats* would have been the ultimate title. (It's amazing how prescient he was. After all, aren't those cat videos on YouTube just soooooo cute? In fact, if aliens looked only at our internet, they could easily think that cats are the dominant life form on our planet.)

So, being the Shallow Fellow I am, I've named this story "Nazis Stole Space Buddha". Yes, this is a deliberately misleading, but hopefully intriguing, title.

Just to put some Hard Fact into this story, this statue is made of iron, stands about 24 centimetres high and weighs about 10.6 kilograms.

NOT EXACTLY BUDDHA

In the interests of full disclosure, I have to confess that when I say "Buddha", I don't mean Siddhartha Gautama, the actual physical man upon whose teachings Buddhism was founded. Rather, I mean a statue. (After all, if the Nazis did "steal" or kidnap the Buddha, who was born around 520 BC, it would imply they can time travel. Of course, any decent conspiracy magazine can tell you that the Nazis have a base on the moon, as well as under Antarctica. This time travel stuff is just nonsense).

To be more correct, the statue was actually not of the Buddha but, more likely, of a very early Buddhist god, Vai'sravana, the Buddhist King of the North, also known in Tibet as Jambhala. (You can see that I'm still telling Porkie-Pies, because Buddhists don't have gods – it's a nontheistic religion.) A solid clue to my deviousness is the statue's right leg. It's hanging down a little, and not mirroring the left leg. In a genuine seated-Buddha statue, both legs should be fully cross-legged.

In fact, laying all my cards on the table, the statue came from a nature-based pre-Buddhist religion called "Bon", based in the western parts of Tibet. However, this religion did get absorbed into Buddhism.

You say Meteor, I say Meteorite

While a rock is zipping through space, it's called a "meteoroid".

Once it enters the Earth's atmosphere, it's now a "meteor" or a "shooting star". (No, it's not actually a star. A real star is thousands or millions of kilometres across.) It starts to flare up and burn around 90 kilometres above the ground.

Meteors get their name from the Greek "meteoros", meaning "lofty" or "high (in the sky)". It's the same root that gave us "meteorology" – which is the complicated interdisciplinary scientific study of the atmosphere.

When the rock has landed on the ground, it's called a "meteorite".

About 40,000 tonnes of meteorites hit our planet each year. (See "Weight of the Earth" in my 34th book, *Game of Knowns.*)

YES, SPACE

While I might have bent the truth a little (OK, a lot) with the mention of the Buddha, I was fairly accurate with the "Space" description.

This statue was made out of iron from a meteorite.

About 15,000 years ago, there was a fall of meteorites along the Chinga stream that ran near the border between Southern Siberia and Mongolia. Some 250 meteorite fragments have been recovered. They range in weight from 85 grams to over 20 kilograms, with a

About 40,000 tonnes of meteoroids hit our planet each year.

total combined weight of over 200 kilograms. About 1000 years ago some unknown people in Western Tibet fashioned this meteorite iron into a statue.

THE IRON AGE

Various human societies worked out how to turn iron ore into iron, and so entered the Iron Age, at different times in different parts of the world. The exact timings are controversial, but you can say roughly 1500 BC in sub-Saharan Africa, around 1200 BC in Europe, India, China and the Ancient Near East, around 400 BC for Korea and 100 BC for Japan. (There are lots of fashioned-iron implements from earlier than these dates in each of these locations – which is why it's controversial!) Iron was much better for weapons and farming implements than stone, copper or bronze.

Early human iron would have been mostly cast iron – hard, but brittle. Meteor iron contains the element nickel. With low nickel levels, meteor iron could be shaped fairly easily by hammering it. At higher nickel levels, you would have to work a lot harder to shape it, but it would quickly become hard enough to be sharpened into a very good weapon.

So before the Iron Age, chunks of meteorite iron that fell out of the sky were highly prized. Indeed, the oldest artefacts made from meteorite iron are some 5000-year-old Egyptian beads. That's thousands of years before the earliest hint of iron smelting in Egypt.

Iron from Heaven

Back before the Iron Age, virtually all iron used by humans came from outer space. (There were some outcrops of Earthly so-called "telluric iron", but they were exceedingly rare.)

Indeed, both the Sumerian and Hittite names for "iron" (*an-bar* and *ku-an*) mean "fire from heaven". Similarly, both the Hebrew and Assyrian names for "iron" (*parzil* and *barzillu*) come from the common root of *barzu-ili*, which means "metal of God" or "metal of heaven". The Egyptian for "iron" is *bia-en-pet*, meaning "thunderbolt".

Meteorite iron is different from Earthly iron. It is usually higher in nickel (5 to 15 per cent), which makes it resistant to rusting. Indeed, when artificial nickel-steel was first produced in 1890, it was marketed as "Meteor Steel". The first modern Stainless Steels appeared in the early 1900s, containing chromium and nickel.

GREENLAND IRON AGE

Greenland had only a very small population and hardly any trees, so it never developed an iron-smelting industry. However, the Greenlanders got their iron from three huge meteorites that all fell within 10 kilometres of each other.

The biggest, named "The Tent", weighed 33 tonnes. "The Woman" weighed three tonnes, while the last, called "The Dog", weighed 430 kilograms. By the time the American explorer Robert Peary went for a look in 1894, the local Inuit had laboriously chipped "The Woman" down to just one and a half tonnes. The anthropologist T.A. Rickard wrote, "The Eskimos obtained battered scraps of the metal by pounding repeatedly in the same place until a small ridge of metal was formed, and then worried it apart . . ."

It was a very slow and tedious task to remove tiny pieces of meteorite iron. It was both tough and malleable – and these were exactly the qualities that made it so valuable. The local Inuit used stones as hammers, but these stones were softer than the meteorite iron – so they had to use lots of them.

How many stones is lots? Well, the pile of discarded stones formed an enormous cone some 6 metres high and 60 metres in circumference.

Meteor Weaponry in Ancient World

The Aztecs of Mexico turned meteorite iron into knives and daggers.

Many devastating conquerors from other cultures (Attila, Timur, Antar, etc.) had swords made from meteorites.

On a smaller scale, Amerigo Vespucci (after whom the Americas were named) said in the 15th century that Indians on the La Plata River in South America used arrowheads made from meteorite iron.

DEFINITELY NAZIS

Okay, that's got "Space Buddha" out of the way – but where do the Nazis come in? Think of the Indiana Jones films.

Carved right in the middle of the chest of this statue made from meteorite iron is a swastika. It is an anticlockwise swastika – the symbol of Auspiciousness, Good Luck and, what the heck, the Female Principle of Divine Energy. This symbol was used in the Indus Valley (on the border of what are now Pakistan and India) at least 4500 years ago, and in many other parts of the Ancient World.

Somehow Heinrich Himmler, the chief of the SS, found out about the swastika on this particular statue. In 1938, he sent an expedition of German scientists to find it, led by the famous zoologist and ethnologist, Ernst Schäfer. Mind you, the Nazi swastika had its arms pointing in the opposite direction (clockwise, not anticlockwise) and had a 45° rotation, but that didn't worry Himmler. He was desperately keen to find any bit of evidence to show that the origins of the "Aryan Master Race" were in Tibet. However, once the statue arrived in Germany, it vanished into a private collection and reappeared only in 2009.

Regardless of its religious value, and its rather convoluted recent history, it's still the only known statue of the human form carved from a meteorite.

Mind you, in this fitness-obsessed world, there are plenty of would-be "super beings" with rock-hard bodies . . .

Meteors and Astrology versus Astromancy

"Astrology" claims that the stars cause certain events. "Astromancy", however, claims that stars merely announce events – events that can sometimes be changed if we are quick enough to carry out the appropriate magical actions.

Astromancy was quite popular in the Middle East some 3000 years ago. The astromancers looked carefully at the motion and characteristics of the incoming meteor. They could range from a "bolide" (extremely bright slow-moving fireball, accompanied by smell and/or sound) all the way down to a barely visible "faint meteor". The motion would include its direction of travel, its appearance time (day or night, early or late), how long it was visible (short or long transit), whether it vanished over the horizon or to your immediate left or right – and so on. These were then given various interpretations – "the enemy will fall in battle", "we will defeat the enemy", "propitious for King's desires", "famine will be in the land", "it will rain thrice in that month", and so on.

I Have a Meteorite

My father was (among his many other careers) a writer. One night, he was staring out of the bedroom window, searching for inspiration. (Before the Web, that's how people got their ideas.)

He saw a shooting star flash past the window.

In the morning, he took me out to look for it in the front yard. Eventually, we found a hole in the ground, about 3 centimetres across. We dug down and found a meteorite – the size of a golf ball.

I still have it.

19

ORGASMS VIA FOOT

The rock musician and composer Frank Zappa once said something like, "you can't have too much sex or vegetables". So, let me ease my way into the simultaneously almost unbelievable and titillating topic of Foot Orgasms.

Yup, foot-contact-caused orgasms. This is the Industrial-Grade version of Footsies.

But, first, I need to tell you about Referred Sensation.

And I'll drop in a Public Health message on top.

HEART ATTACK 101

I'm starting off with heart attacks to explain Referred Sensation.

You might have seen (on TV and in movies) that heart attacks often cause a pain in the left arm or shoulder. This is despite the fact that the heart sits in the chest.

The heart is a hollow muscular organ that pumps about five litres of blood each minute. In a heart attack, the problem is that the muscles doing the pumping don't get enough oxygen. These muscles complain – surprisingly, not just as a pain in the heart, but sometimes as a pain felt in the left arm or shoulder.

Why is the pain not always felt in the heart?

Pain felt away from its source is classic Referred Sensation. The pain (that is, a sensation) coming from an internal organ (the heart) is shifted or "referred" to an external location – the left arm or shoulder.

Public Health Message

A heart attack (or myocardial infarction) is a genuine and serious medical emergency. If somebody ever has symptoms suggesting a heart attack, they (or you) should ring the Emergency Medical Services immediately on 000.

The most common symptom of a heart attack is chest pain. It's often described as pressure, squeezing or tightness. The pain most commonly radiates to the left arm, but can also travel to the right arm, lower jaw and neck, back and upper abdomen. In the upper abdomen, it can mimic heartburn.

Other symptoms can include excessive sweating, nausea and vomiting, weakness, palpitations of the heart, and feeling light-headed.

Trickily, about one quarter of heart attacks are "silent" – they have no symptoms. They are found via blood tests or ECG, or at autopsy.

I have spoken to people who knew their symptoms were because of a heart attack, but they waited before going to hospital. One person deliberately delayed and went to hospital after hours, because parking would be so much easier. Unfortunately, his outcome was worse than if he had gone immediately.

DON'T WAIT – RING 000 IMMEDIATELY!

REFERRED SENSATION 101

We don't really understand the mechanism by which Referred Sensation happens, but one reasonable theory runs like this.

The heart sends sensory information to your brain, via some "relay stations" in the nerve centres at the top of your spinal cord. By a coincidence, your left arm and shoulder use those very same relay stations to send sensory information to the brain.

Decade after decade, your heart works within its limits. It never complains. Over a lifetime of trouble-free pumping, your brain doesn't consider the heart. It gets used to the idea that any sensations coming from those relay stations at the top of your spinal cord are coming *only* from your left arm.

But one day (and, as they say in *The Godfather*, let's hope that day never comes), you have a heart attack. This is potentially life-threatening. So your heart complains, and sends nerve signals towards your brain. This is the very first time in your whole life that your heart has sent this kind of nerve signal to your brain. Your brain sticks to what it knows. You interpret this unfamiliar sensation of a heart attack as pain in your left arm.

FOOT ORGASM = "REVERSE REFERRED SENSATION"

In the Referred Sensation of a "heart attack", signals from an internal organ (the heart) are referred to an external location.

It's the opposite for Foot Orgasm. It is a kind of Reverse Referred Sensation.

In the Referred Sensation of a "Foot Orgasm", signals from an external location (the foot) are referred to an internal organ (the genitals).

The Foot Orgasm Syndrome was first described in 2013 by Marcel D. Waldinger. (In fact, there's only one case in the entire medical literature.)

Let's start with some basic neuro-anatomy. Both the foot and the genitals share the same relay stations down at the bottom of your spinal cord. (You can see where I'm heading.)

The brain of a 55-year-old woman wrongly interpreted sensation from her left foot as coming from the internal genital area. In her specific and very uncommon situation, sensation on the bottom of her left foot would lead to an orgasm.

The brain of a 55-year-old woman wrongly interpreted sensation from her left foot as coming from the internal genital area.

The result was spontaneous, unwanted and, yes, inconvenient, orgasms – with absolutely no sexual desire or arousal. They were different from the woman's regular orgasms. These Foot Orgasms occurred very suddenly without any pre-orgasmic build-up, were very short (lasting only 5 to 6 seconds) and finished in a very uncharacteristic and abrupt manner. She would need a few minutes to recover. They happened five or six times each day, often accompanied by vaginal lubrication.

Foot As Erotic Symbol

Different people see the foot in "different" ways.

"Foot Fetishism" happens when one's sexual arousal is dependent on looking at, or caressing, someone's foot or feet.

It is claimed (with perhaps not too much evidence) that the shape of the foot in a high-heeled shoe is similar to the shape of the foot in (female) orgasm. Indeed, Dr Waldinger described this as "when feet and toes may automatically go into plantar flexion resulting in arching of the foot and curling of the toes". (I couldn't find any comments in the medical literature about what happens to the male foot.)

FOOT STORY BACKSTORY

The story began in 2008, when this woman had an operation on her gall bladder. There were some unfortunate complications, and she became very sick indeed. As a result of massive infection and breathing problems, she had to stay in Intensive Care and the

Department of Surgery for three weeks. The medicos reckon that at this stage, the relay stations at the bottom of her spinal cord were damaged. As a result of the damage, she had some strange tingling and burning sensations in her left foot.

One and a half years later, in June 2010, she was prescribed a SSRI (Selective Serotonin Re-uptake Inhibitor) anti-depressant, Paroxetine, to try to relieve the abnormal foot sensations. Anti-depressants are used in low doses to modify experiences of pain.

In the case of our 55-year-old woman, within a few days of taking Paroxetine, the Foot Orgasms began. We don't fully understand the mechanism. Perhaps taking a powerful serotonin-acting drug, at the same time as the nerve endings in her damaged nerve roots were still trying to heal, caused some kind of cross-wiring. This cross-wiring was permanent.

DIAGNOSIS AND TREATMENT

About one year later, the woman sought medical treatment. This was a difficult decision for her. At her first medical appointment, she was deeply embarrassed and crying uncontrollably.

This was a complicated case, so the work-up included not just a physical examination, but Sensory Testing, MRI scanning, ElectroMyoGraphy and Trancutaneous Nerve Stimulation.

After a lot of work, the medicos tried a procedure that was both diagnostic and therapeutic. They injected anaesthetic into the left S1 dorsal root ganglion at the base of her spinal cord. This reduced the frequency of the Foot Orgasms by 50 per cent, and their strength by 80 per cent. And when the doctors went further, and combined the anaesthetic with Pulsed Radio-Frequency treatment, the Foot Orgasms stopped entirely.

So if you start experiencing unwanted and inconvenient orgasms, perhaps you should tread carefully . . .

20

PALEOLITHIC DIET

Today there are many diets available to anybody wanting to lose some weight – some might say too many. One current favourite is the so-called "Paleolithic Diet".

In a nutshell, it claims that our Paleolithic hunter–gatherer ancestors lived in wondrous harmony with their environment, which gave them exactly the food they needed to live a long and happy life. This sweet congruence came to an unhappy end with the development of agriculture and grain-based diets. Furthermore, the Paleo Diet claims that the period of 12,000 years or so since we invented agriculture is far too short for our bodies to have evolved to cope with the new foods that agriculture has given us.

In other words, the key to a healthy and long life is to abandon our modern agricultural diets and eat what our Paleolithic ancestors ate.

But this claim is false – and has no basis in dietetic, evolutionary or archaeological reality.

What Is the Paleo Diet?

The Paleo Diet is also called the Stone Age Diet, and the Hunter-Gatherer Diet.

It was first promoted by the gastroenterologist Walter L. Voegtlin in the mid-1970s. He argued that we humans are carnivores (wrong), and that our Paleolithic ancestors ate a carnivore's diet (wrong again). Therefore, he claimed, our diet should be meat and fat, with tiny amounts of carbohydrates.

The Diet was re-invented by S. Boyd Eaton and Melvin Konner in 1985. They called it the "Evolutionary Discordance Hypothesis".

This Diet is based on grass-fed, pasture-raised meat and poultry, as well as seafood, along with some fruit, green vegetables, eggs, nuts, roots and fungi. Sounds quite reasonable.

It also advocates excluding dairy, grains, legumes, potatoes or processed oils. Less reasonable – I love them all, especially olive oil!

It also excludes refined salt and refined sugar. It's an excellent idea to minimise consuming lots of refined sugar, but what about the occasional birthday cake?

PALEO PROBLEMS

The Paleo Diet has major problems at every possible level – from theoretical to practical.

First, our ancestors ate many very different varieties of Paleolithic Diets. There was no one single Paleo Diet for all the humans across our planet.

Second, we humans actually have done a lot of evolving in the last 12,000 years. That includes evolution in regard to what we can eat.

Third, we can't eat what our Paleolithic ancestors ate anyway – because most of that stuff is not around any more.

And fourth, the recommended Paleolithic Diet is way out of kilter with what dieticians currently recommend.

PROBLEM 1: JUST *ONE* PALEO DIET?

"Paleolithic" literally means "Stone Age". The Paleolithic era spans a period from around 2.5 million years ago, right up until the development of agriculture some 12,000 years ago. (I discuss Agriculture in my 17th book, *Flying Lasers, Robofish and Cities of Slime*.)

When people started writing books about the Paleolithic Diet back in the 1970s, we had only a very vague idea of what our Stone Age ancestors ate. (Yup, the original writers advocating Paleolithic Diets were far more like "creative writers" than "factual writers".) We did have hints of a bias towards a diet focused on meat. For example, we have found paintings, some 17,300 years old, inside the Lascaux Cave in the Dordogne region of France. They show animals, and people hunting animals. They don't show any agricultural fields.

But since then our anthropologists and archaeologists have looked at fireplaces, middens (dunghills or refuse tips), the actual teeth of our Paleolithic ancestors, and even the tools used to prepare their meals. It turns out that they ate a highly varied diet. Cereals and grains are forbidden in the Paleolithic Diet. But we know for sure that our ancestors ate them. The evidence for this came from examining dental plaque and wear marks on their teeth, as well as the tools they used to process food.

Our bodies could easily evolve fast enough in 12,000 years to accommodate new foods. In fact, they have.

And was there one single Paleolithic Diet, right across the planet? Did everyone eat the same meal in what we now call Africa and Europe, the Americas, Asia and Australia? "No" from common sense, and "no" from what our archaeologists and anthropologists have found.

We see an incredible spread in the diets of some of the so-called "primitive" peoples. The Inuit of the Arctic get 99 per cent of their calories from meat, while the !Kung people of Africa eat around 12 per cent meat. That's a huge range. They definitely don't eat the same identical meal.

What Do Dieticians Say?

A 2011 report involving 22 experts rated 20 diets. Based on factors such as health, ease of following and weight loss, the Paleo Diet came last. In 2012, it tied for last place with the Dukan Diet for the lowest rating – 29th out of 29 diets.

To complete its losing streak, in 2014, Paleo tied for last place (32nd out of 32) with the Dukan Diet.

PROBLEM 2: COULD NOT EVOLVE FAST ENOUGH?

Another cornerstone of the Paleolithic Diet Creed is that our bodies could not possibly have evolved fast enough in the last 12,000 years to accommodate our new foods. Indeed, the promulgators of the Diet claim that our genes haven't changed for 50,000 years.

This is so very wrong. Evolution can be quite quick on the uptake.

We have very solid evidence (for example, from their teeth) that 30,000 years ago some of our ancestors were already eating grains and legumes.

In the last 7000 years, about one third of us have evolved to be able to drink milk when we grow into adults (see story on "Lactose Intolerance" on page 151). There are 6000-year-old rock paintings of people herding domestic cattle in the Jebel Acacus region of the Sahara Desert in Libya. Seven thousand years is definitely less than 12,000 years!

Still concentrating on food, some of us have evolved extra copies of the amylase enzyme so that we can more easily digest starches. Furthermore, some Japanese have evolved special bacteria in their guts that can digest seaweed – so sushi is no trouble at all. (See "The Stranger Within" in my 31st book, *Brain Food*.)

Moving away from evolution related to food, some of us have evolved blue eyes (6000 to 10,000 years ago). Others among us, in Africa, evolved resistance to malaria (5000 to 10,000 years ago).

Consider the challenge of living at high altitudes. Three separate groups of humans living in Tibet, the Andes and in Ethiopia have evolved three different methods of dealing with low oxygen.

So, yes, our bodies could easily evolve fast enough in 12,000 years to accommodate new foods. In fact, they have.

Weight Loss?

One of the claims of the Paleo Diet is that it triggers the production of hormones that then suppress hunger. In turn, this would produce the desired weight loss – the whole point of any Fad Diet.

Researchers found this to be incorrect.

The hunger-suppressing hormones are not triggered by the diet.

PROBLEM 3: EAT WHAT OUR ANCESTORS ATE?

The third problem with the Paleolithic Diet is that the food eaten back then is simply not around any more. We have transformed the meat and plant species we eat through millennia of artificial selection and evolution.

If you look at what comes from today's food animals, very few meats are as lean as those our Paleolithic ancestors ate. Indeed, many of the larger food animals have gone. There are no more mammoths or moas, and the last auroch (a super-large cow) died in Poland in 1627. However, kangaroo meat is pretty lean.

Today's corn started off as a straggly skinny grass in Central America, while tomatoes used to be tiny berries. Bananas were mostly filled with seeds until a recent mutation (discussed in my 24th book, *Disinformation*). Consider cabbage, broccoli, Brussels sprouts, cauliflower and kale: they might look wildly different today, but they are each cultivars of one single species, *Brassica oleracea*.

Modern versions of the Paleo Diet recognise that foods have changed, and allow domesticated animal meat and cultivated plants.

S. Boyd Eaton and Melvin Konner have recently "adjusted" their 1985 version of the Paleo Diet to allow whole grains and low-fat dairy products.

PROBLEM 4: NUTRITIONAL VALUE?

The fourth problem with the Paleolithic Diet is its nutritional aspects.

The core recommendation of the Diet is a high protein intake – 19 to 35 per cent of a person's daily energy. This is quite a lot higher than the Australian Nutrient Reference Values suggestion of 15 to 25 per cent. Indeed, diets rich in meat are associated with higher rates of heart disease. The Paleolithic Diet also recommends a moderate to high intake of fat – again, not recommended by modern dieticians.

The Diet advises not to eat any whole grains. However, we have very solid evidence (for example, from their teeth) that 30,000 years ago some of our ancestors were already eating grains and legumes.

But, on the plus side, the Paleolithic Diet advises against eating processed foods with added salt, sugars and flavourings – entirely sensible. It also recommends fibre from vegetables and fruit – an excellent suggestion.

OUR GUT IS GREAT

The Paleo Diet relies on the underlying fantasy that, if we simply follow it, we'll change from a balding, pot-bellied man slouched in front of a computer into a tall, well-muscled man with perfect teeth, an artfully placed fur loincloth and a spear.

Sure, sitting all day is not good for you. (I recently changed over to a desk that adjusts from standing to sitting.)

But our gut is perfectly adequate for many different diets. After all, meat eaters, vegetarians and vegans can all be very healthy.

Our digestive system has adapted to eat most foods. Our mouths are equipped with the teeth of both carnivores and herbivores – we can tear meat with our canines, and we can grind fibrous plants with our molars. The gut that runs between our mouth and anus is totally different from a straight line (the shortest distance between two points). Instead, it's about 10 metres long. It's not the short gut of a carnivore. Neither does it have the multi-stomach fermentation chambers of a grass-eating herbivore. It's in between.

Gregor Yanega, a professor of biology at Pacific University in Oregon, has said, "Our guts are special because they are less specialised. They can accommodate so many changes in the foods that surround us, can accommodate unusual abundance and a certain amount of scarcity: we can even eat some of the world's more difficult foodstuffs: grains, leaves and plants. Berries, nuts, meats, sugars, those are easy. Eating them together is pretty rare."

Maybe we should forget Fad Diets, and just remember Michael Pollan's simpler and more useful advice: "Eat food. Not too much. Mostly plants."

21

PERMAFROST FEEDBACK LOOP

The Arctic is getting a bit of a hiding from humanity. The volume of the floating summer ice in the Arctic has been pretty constant for at least 1400 years. But since 1980, its volume has dropped by 80 per cent. (Check out "Arctic Meltdown – Milankovitch Cycles "on page 53.)

Now it's become obvious that we are affecting another part of the Arctic – we are thawing out its permafrost and releasing the Greenhouse Gases contained therein. The real worry here is that we might be switching on a Positive Feedback Loop that we can't switch off.

PERMAFROST 101

Permafrost is defined as any ground that has been frozen for at least two years.

Most of us don't realise just how much permafrost there is. I was astonished to learn that about one quarter of all the land mass in the Northern Hemisphere is permafrost. Permafrost makes up 85 per cent of the surface area of Alaska and 55 per cent of Russia and China's surface area.

Once the average annual air temperature drops below 0°C, some of the ground that freezes in winter will not completely thaw out the next summer. Providing the air temperature stays low enough, a layer of permafrost forms at the surface – and gradually grows downwards. Permafrost can be up to 1500 metres deep in northern Siberia, and 740 metres deep in northern Alaska.

What's known as the "active layer" of the permafrost is a thin surface layer that thaws each summer and refreezes each winter. It ranges from less than 30 centimetres thick in the continuous permafrost on the Arctic coast to more than 2 metres thick in the discontinuous permafrost of southern Siberia. Plant roots cannot push through permanent permafrost, so vegetation can survive only in the active layer.

Permafrost Pedantics

Permafrost is less common on south-facing slopes that receive more solar energy (this holds for the Northern Hemisphere). It's also less common where there is an insulating medium such as vegetation or snow.

Permafrost is classified as continuous when it covers 90 to 100 per cent of the land area, discontinuous when it covers 50 to 90 per cent, sporadic when the land covered is 10 to 50 per cent, and isolated when less than 10 per cent of the land is permafrost.

THAWING PERMAFROST

But thanks to the recent warming up of the permafrost, there are now vast areas covered by so-called "drunken forests". Here, the permafrost has warmed and so the ground has softened. As a result, the trees have fallen until they have landed on other "drunken" leaning tree trunks.

The volume of the floating summer ice in the Arctic has been pretty constant for at least 1400 years. But since 1980, its volume has dropped by 80 per cent.

This is bad enough, but the real worry relates to the potential release of Greenhouse Gases that would follow the thawing and then melting of the permafrost.

There are enormous quantities of Greenhouse Gases such as carbon dioxide (CO_2) and methane (CH_4) encased in the frozen permafrost.

By "enormous", I mean about 1.7 trillion tonnes of organic carbon stored in the permafrost. This carbon comes from the remains of plants and animals that have accumulated over many many thousands of years. The carbon locked up in the permafrost is roughly equal to four times as much carbon as we humans have dumped into the atmosphere in modern times – or twice as much carbon as is in the atmosphere right now.

Recent Changes

Previously frozen lakes have begun bubbling constantly with uprising methane – which can be ignited to make a spectacular fiery plume. (Check out the videos on YouTube.) Widespread giant plumes of methane gas have begun bubbling up from the floor of the Arctic Sea – when previously they didn't.

TODAY'S SITUATION

This current cycle of permafrost probably began with the current series of Ice Ages, about 3 million years ago.

Even though the permafrost covers about one quarter of the land mass of the Northern Hemisphere, we have only two global networks to monitor its status.

> The warming atmosphere first thaws the permafrost. The carbon dioxide released is bad enough. But methane is some 22 times more active than carbon dioxide as a Greenhouse Gas. This release of carbon dioxide and methane warms the atmosphere even further. This then thaws more permafrost.

One group, the Thermal State of Permafrost network, measures the temperature at various depths at 860 locations. It tells us that permafrost temperatures have climbed over the past few decades.

The other monitoring group, the Circumpolar Active Layer Monitoring network, has only 260 sites where it measures the thickness of the active layer of the permafrost. This network tells us that the active layer of the permafrost has been getting thicker – in other words, the warmth is migrating downwards.

Costs

The potential costs to the world economy of Global Warming are stupendous.

Gail Whiteman, Professor of Sustainability, Management and Climate Change at Erasmus University writes about "the release of methane from thawing permafrost beneath the East Siberian Sea, off Northern Russia". Methane release in that area alone comes with "an average global price tag of $60 trillion . . . comparable to the size of the world economy in 2012 (about $70 trillion)".

There is a sad similarity between global warming and cigarettes. The revenue is gathered today, the costs get shifted to tomorrow. This kind of thinking is on a par with, "I have milk in the fridge. Why should I worry about cows?"

POSITIVE FEEDBACK LOOP

The real worry is the possibility of a "Positive Feedback Loop". (I discuss Feedback Loops in "Gold in Trees" on page 115.)

In this terrifying scenario, the warming atmosphere first thaws the permafrost. The carbon dioxide released is bad enough. But methane is some 22 times more active than carbon dioxide as a Greenhouse Gas. This release of carbon dioxide and methane warms the atmosphere even further. This then thaws more permafrost. That then releases more Greenhouse Gases – and on, and on, and on, repeatedly.

Once this scenario happens, we humans no longer have to dump any more Greenhouse Gases into the atmosphere. The permafrost, the warming planet, and the Greenhouse Gases are in their own

"Positive Feedback Loop". In fact, the warming–gases–warming cycle would still proceed, even if all the humans on Earth left the planet and went to Mars. The Positive Feedback Loop of warming–gases–warming stops only when the Greenhouse Gases have all been freed from their frozen prison in the permafrost.

Will this scenario occur?

Well, we will know with 100 per cent certainty only when it is too late – only after it has already happened.

But today's Science tells us that if we don't cut back on burning carbon, it looks extremely likely that we are heading into this Positive Feedback Loop. And that's troubling.

CRACK!

22

SELFIE

We Australians are pretty easy to recognise by our turn of phrase – especially the way that we shorten words and then shove a vowel or two on the end. Fire officer becomes "firie", tradesperson becomes "tradie", and a tin of beer gets called a "tinnie". So lend me your "earies", and I'll tell you the story of how "self-portrait photograph" became "selfie". Yep, we Australians brought this new word into the English language – and I had a small part in this process.

Selfie = New Genre?

There are claims that with selfies, art has entered a brand new genre. Supposedly the instantaneous self-portrait, shared via social media to the "audience", is inviting them to connect – and this is a "first" in human history.

Perhaps.

However, millennia ago the Ancient Greeks had the concept of "methexis" in their theatre – where the speaker on stage would turn and speak directly to the audience. In return, the audience would join in with the stage action.

HISTORY OF THE SELFIE

Possibly the earliest selfie was painted by Parmigianino back in 1523 when he created *Self-Portrait in a Convex Mirror.*

We think that the earliest photographic selfie was taken by Robert Cornelius in 1839, in Philadelphia. He was an amateur chemist and photographer. He put the film in the camera, removed the lens cap, sat back and remained perfectly still for one minute, and then reached forward and replaced the lens cap. He then had to process the film with chemicals and dry it before he could look at the first photographic selfie. (There was nothing instant about photography in those days.)

Since then, things have changed. We've even had selfies sent back to Earth from Mars by the two Mars Rovers still exploring there – our first robot selfies from outer space.

For a brief while, the most famous selfie was the one taken at the memorial service of Nelson Mandela, one of the truly inspirational figures of the 20th century. This group selfie had just three people in it: Danish Prime Minister Helle Thorning-Schmidt was in the centre taking the selfie, and was flanked by the British Prime Minister, David Cameron, and the President of the United States, Barack Obama.

But then Hollywood got involved. The selfie orchestrated by Ellen DeGeneres at the Oscars in February 2014 is now probably the best-known selfie. She started off with Meryl Streep and Julia Roberts – and was then joined by Brad Pitt, Angelina Jolie, Jared Leto, Bradley Cooper, Kevin Spacey and more. You know that they must be movie stars, because otherwise they wouldn't be thin enough to all fit into the one picture.

Mars Rover Selfie

Both of the currently operating Rovers on Mars (the solar-powered *Opportunity* and the nuclear-powered *Curiosity*) have sent back selfies. In each case, they used a camera mounted on the end of an arm, took multiple overlapping selfies, and then combined then into a single selfie – minus the arm.

***Opportunity* sent back a selfie on 15 April 2014, to celebrate its sixth Martian winter. It landed on Mars on 25 January 2004, with a three-month guarantee – over 10 Earth years ago. Thanks to recent winds, its solar cells are now the most dust-free they have ever been since its first Martian winter in 2005. On 28 July 2014, it gained the record for the longest distance driven "off-world" – 40 kilometres.**

***Curiosity* sent back its selfie on 24 June 2014, to celebrate the Marsiversary of landing on Mars on 5 August 2012.**

Marsiversary? The year on Mars lasts 687 Earth days.

I wonder how long it'll be before "Marsiversary" enters the Oxford English Dictionary. Probably when we have humans living on Mars . . .

ETYMOLOGY OF "SELFIE"

In the study of language, called etymology, the origin of a new word is defined by when it first appeared in print. In the old days, that meant paper, but times have changed.

The word "selfie" was officially born back in 2002, with Nathan Hope, hereafter known as Hopey – so you know that he's Australian. He went out for a mate's 21st birthday and had a little mishap. On 13 September 2002 at 2.55 p.m., he went onto an online forum (as Hopey) to ask about the dissolvable stitches that were by then in his lower lip. They were dry and uncomfortable. After a bit of chat back and forth, the entity known as "My Evil Twin, Beryl" asked him how he came to get these stitches, and at 3.19 p.m., he typed in reply:

"Um, drunk at a mate's 21st, I tripped ofer [sic] and landed lip first (with front teeth coming a very close second) on a set of steps. I had a hole about 1 cm long right through my bottom lip."

He then posted a "self-photograph" showing the stitches in his lower lip.

He continued writing: "And sorry about the focus, it was a selfie."

That was the very first written use of the word "selfie", in any medium (paper or electronic).

That's how the word "selfie" got into the English language. Its use ramped up slowly and steadily for a while, but then took off massively in 2013. In November that year, it was declared that over the previous year, the usage of "selfie" had increased by an astonishing 17,000 per cent. Who said so? The *Oxford English Dictionary* – or OED

– which, by the way, is *the* definitive record of our rapidly evolving English language.

How did the OED measure this? Well, they analysed the Oxford English New Monitor Corpus, which is an archive of electronically stored structured sets of texts – fancy talk for "whatever we could grab off the web, and store, and file". Each month some 150 million words are collected. This database is statistically analysed every day to track new and emerging words – and "selfie" began to stand out from the pack.

> The word "selfie" was officially born back in 2002.

So, in November 2013, the OED declared that for 2013 the Word of the Year would be "selfie".

And in what forum did Hopey post the first-known use of "selfie"? It was my very own "Dr Karl Self-Serve Science Forum" on the ABC! And yes, I did Tweet about that . . .

Really Cool?

The word "selfie" was added to the *Oxford English Dictionary* in August 2013, in their quarterly update of new words.

But in the Land of Etymology, there is another specialised, elite dictionary, jam-packed full of ridiculous words. On 5 August 2014, "selfie" jumped into *The Official Scrabble Players Dictionary.*

Potential Words of the Year, 2013

There were several runners-up for the Word of the Year in 2013. They included "twerk" (to dance in a low squat in a sexually provocative manner, using thrusting hip movements), "binge-watch" (to watch many episodes of a TV show in one bout) and "showrooming" (to visit bricks-and-mortar stores to examine merchandise before buying it online).

23

SLEEP, MYSTERIOUS SLEEP

It seems reasonable that we don't understand difficult and complicated stuff like Quantum Mechanics and Dark Energy. But how come we also don't understand "simple" stuff, such as why we sleep? Why do we spend about one-third of our lives virtually paralysed, with our eyes shut?

Sure, we sleep because we are sleepy. But why do we get sleepy? And why after about 16 hours of being awake? Why not one hour, or 60 hours, after waking up?

WHAT HAPPENS IN SLEEP

It's not that the brain is resting while you sleep. In fact, the opposite happens. When you sleep, networks in the brain undergo very organised patterns of activity – and use almost as much energy as when you are awake.

We also know that during sleep, the brain orchestrates the manufacture of many different types of molecules. These include proteins, steroids, cholesterol, lipid rafts, human growth hormone, and many more. So sleep has benefits. If you don't get enough sleep (as happens to many shift workers), you suffer increased risk of heart disease and some cancers.

But why do we sleep? Almost certainly, it'll turn out that there are many different reasons – all of them correct to varying degrees.

Fun Question

It's not often that you get to run into a Sleep Scientist. I was lucky enough to do this recently.

I was finally able to ask the question that I had wanted to ask for about 20 years. (I actually knew what the answer would be, but I wanted to hear it from a genuine Sleep Scientist.)

"Can I ask you this? Why do we sleep?"

Her straightforward and straight-faced reply was, "We really don't know."

Try it next time you run into a Sleep Scientist.

BASIC THEORIES OF SLEEP

Some Sleep Scientists claim the ultimate purpose of sleep is to "prune" or weaken brain connections that you have made while awake. This means that the next day, your brain won't be overloaded with irrelevant memories. But, supporting a completely opposing point of view, other Sleep Scientists have done research that suggests that sleep replays and consolidates significant memories.

> When you sleep, networks in the brain use almost as much energy as when you are awake.

Other recent studies point out that during sleep, some nerves send their electrical impulses in the opposite direction to what they normally do when you are awake. We're not sure why they do this.

Many Pacemakers

Over the last century, we have slowly and painstakingly found circadian pacemakers (such as the Supra-Chiasmatic Nucleus or SCN) that control various rhythms in the body. These rhythms include body temperature, blood pressure, how quickly your reflexes react – and, yes, sleep.

These internal body rhythms slide out of synchronisation with the external cycles of day and night when you travel, causing jet-lag.

By the way, when you change time zones, the SCN can shift itself by only one hour per day. Confusingly, some of the organs it controls (body fat, liver, lung, heart, and so on) can shift by more than one hour per day.

THE LATEST THEORY

> Sleep gives the brain the opportunity to flush out the various metabolic waste products created during the day.

The most recent research offers yet another alternative theory as to why we sleep. It says that sleep gives the brain the opportunity to flush out the various metabolic waste products created during the day.

If you look at the anatomy of the brain, there's a lot of obvious stuff – the nerve cells, the glial cells that support the nerve cells, the arteries and veins, and the Cerebro-Spinal Fluid (or CSF) that keeps the brain afloat.

What's under appreciated is the gap in between these structures – the so-called "Interstitial Space" – that is filled with the wondrous liquid called CSF. This Interstitial Space makes up about 20 per cent of the volume inside your skull.

The brain weighs about one and a half kilograms. The trouble is that it has virtually no structural integrity – it's all soft and squishy. Without CSF, the weight of the brain would crush all the structures that happen to be at the bottom of the brain. So one very important role of the CSF is to buoy up, or float, the brain so that its effective weight is only about 25 grams – and the stuff at the bottom of the brain doesn't get damaged by the weight of the stuff above it.

At any given moment, there's about 140 millilitres of CSF in and around the brain, and running up and down the spinal cord. However, you make about 500 millilitres of CSF each day, so it's always being created – and always being absorbed. This means that you have massive over-capacity making extra CSF. Most of this CSF flows through the Interstitial Space.

During sleep, the brain cells shrink, and so the Interstitial Space increases by about 60 per cent in volume. This means that when you sleep, the flow rate of the CSF in and around your brain increases by an astonishing 20 times. Suddenly, the waste products of brain metabolism that have built up during the day are whisked away by the raging torrent of Cerebro-Spinal Fluid.

Poiseuille's Law

Consider water flowing through a pipe. You would think that if you doubled the diameter, you would double the flow – the volume of water that flows through each second.

But no. If you double the diameter, the volume of water that can flow through each second is 16 times more.

Poiseuille's Law tells us that the flow is related to the Fourth Power of the diameter. The law was derived in 1838.

In the case of your heart, consider what happens when one of the coronary arteries has its diameter halved (perhaps due to internal plaque clogging it up). The flow of blood drops not by a factor of 2, but by 16. Instead of having 100 per cent of your blood flow through that artery, you now have only 6 per cent. That's not enough to supply your heart muscle, and you have a heart attack.

GO WITH, OR AGAINST, THE FLOW?

We know that in some cases, excess quantities of proteins in the brain are linked to various neurodegenerative diseases. For example, the protein ß-amyloid is linked to Alzheimer's Disease. This, and many other similar proteins, are normally present in the Interstitial Space.

Scientists have carried out a rather specific research study. They deliberately administered drugs to stop the flushing action of the massively increased flow rate of the CSF, which happens during sleep. As a direct result, the levels of nasty waste-product chemicals in the brain increased enormously.

Perhaps new drugs could increase the flushing action of the CSF even while we are awake. If we could reduce the levels of these nasty chemicals, maybe then we could get de-wasted in the process . . .

24

SMARTER THAN YOUR PARENTS, NOT AS SMART AS YOUR KIDS

We've all heard of IQ Tests. (IQ stands for Intelligence Quotient, if you were wondering.) They are generally viewed as being a helpful, yet imperfect, measure of how "smart" we are. The Psychologists see IQ as our "fluid and crystallised intelligence" – our ability to solve problems, and to reason. It's supposedly a measure of our ability to interpret, process and manipulate information – both deeply and speedily.

But surprisingly, averaged around the world, the typical IQ is not fixed. IQ seems to be creeping upwards, at about three IQ points every decade.

IQ Changes

Your IQ is related both to your genetics (which you got from your parents) and to your environment.

Consider the changes that can happen when a poor child is adopted into a wealthier family. Their measured IQ can increase by 12 to 18 IQ points.

Or consider a study that measured the IQ of adolescents on two occasions, four years apart. Some of them changed their results by 20 IQ points.

THE HISTORY OF THE FLYNN EFFECT

This story of rising IQs begins back in 1948. That was when the psychologist R.D. Tuddenham examined the IQ scores of American men who had been conscripted into the military between the years 1917 and 1943. He showed that their IQ scores were increasing by about 4.4 points every decade.

In 1949, the Scottish Council for Research in Education showed a similar increase in intelligence over the years 1932 to 1949. This strange increase in IQ over the decades was rediscovered by the psychologist Richard Lynn in 1983, and again by the moral philosopher James R. Flynn in 1984.

Today it's known as the Flynn Effect. (Flynn is now Emeritus Professor of Political Studies at the University of Otago in New Zealand.)

What Do IQ Tests Actually Measure?

The best answer is, "We are not sure."

It seems that IQ tests are a little flawed in their attempts to measure "intelligence". Perhaps, as Flynn wrote in 1987, "they seem to correlate with a weak causal link to intelligence".

IQ TESTS 101

IQ tests are limited. For example, they do not measure musical or artistic creativity, or the so-called Emotional Quotient. They try to measure "something else".

You can see this by looking at a typical IQ test, the Wechsler Intelligence Scale for Children (or WISC). The WISC measures 10 separate cognitive skills. They each try to measure a different "aspect" of this strange beast we call IQ.

These 10 categories include Information ("On what continent is India?"), Arithmetic ("If three toys cost four dollars, how much do seven toys cost?"), Vocabulary (the words we use in everyday life), Comprehension ("Why are houses in a street given numbers?"), Picture Completion (where you find the missing part in a picture) and Similarities ("How are dogs and rabbits related?").

IQ tests are limited. For example, they do not measure musical or artistic creativity, or the so-called Emotional Quotient.

Another widely used test is Raven's Progressive Matrices (RPM). It's a non-verbal and non-mathematical test, suitable for children aged five and up. It uses patterns in an attempt to measure on-the-spot problem solving ability. There are 60 multiple-choice questions, which get progressively more difficult. In each question, you are asked to point out the missing element to complete a given pattern.

RPM seems to be largely independent of the culture you live in, or the education you've had. It gives similar results for Kalahari Bushmen and Eskimos living in the Arctic.

IQs Are "Adjusted"?

You might not have realised this, but every time a new IQ test is introduced, or an old one is updated, the scores that people achieve are mathematically adjusted. This is to give the adjusted scores a so-called "Normal" or "Gaussian" Distribution. (Look it up on Wikipedia.)

The mathematical adjustment starts once the results of the new or updated test have been gathered from a large group of people. Suppose that the average score is 78, and that two thirds of the scores lie between 58 and 88. The scores then get adjusted (or scaled) so the average IQ on this test will be 100, and also to make sure that about two-thirds of people will lie in the IQ band between 85 and 115. The process is called "norming". Each new or updated IQ test will be "normed".

The purpose of this is to make different IQ tests comparable. The average IQ will always be 100.

IQ RISING

So let's get back to the widely used Wechsler Intelligence Scale for Children (or WISC). Over the last two thirds of a century, it has had to be recalibrated three times. Why?

To make sure that the average measured IQ of children was always 100.

The original WISC was released in 1947. It was recalibrated (or "renormed") upward in the early 1970s and renamed the WISC-R, and renormed up again in the late 1980s and called the WISC-III. Most recently it was renormed up in the early 2000s, and renamed the WISC-IV. Each version was harder than the one before it. (And then the WISC-V, the WISC-VI, etc.)

IQ is rising at 3 points per decade, 9 points per generation (assuming 30 years for a generation), and so we can extrapolate to 33 points per century.

Here's the essence of the Flynn Effect. If you give a child of today one of the WISC IQ tests of the past, on average, that child will score more than 100.

But if you look at each of the 10 components that make up the WISC, you'll see differences in their scores over time.

On one hand, there has been virtually no increase in the Arithmetic scores over the decades.

But over the same time, there has been a huge measured increase in the Similarities scores. Why? We're not sure. Our society and our intellectual environment have changed enormously over the last century, transitioning from "concrete" to "abstract". So a century ago, the answer to "How are dogs and rabbits related?" would have been the rather concrete "You use dogs to hunt rabbits". But today the typical answer is the more abstract "Both dogs and rabbits are mammals".

A similar increase has been seen in the non-mathematical and non-verbal Raven's Progressive Matrices. Indeed, the RPM shows some of the highest gains in measured IQ.

THE CAUSE OF THE FLYNN EFFECT

What's causing the Flynn Effect?

We don't know, but many explanations are offered.

> IQ is a measure of how well we can deal with the current society we inhabit. In other words, perhaps it's a measure of how "modern" we are.

First, compared to a century ago, our brains have to work within an environment that is more abstract. For example, today's world is loaded with synthetic visual imagery – from televisions, computers and video games.

One century ago, a phone was only a phone. It certainly was not also a camera, a speaking Spanish–English translator, maps of the entire world, a local bus and train timetable, newspapers, an instant currency converter, a guide to Solar Activity to help you find auroras, a compass, a music player and voice recorder, and a personal exercise trainer.

According to James Flynn, back in 1900 only 3 per cent of Americans worked in jobs that were "cognitively demanding". Most jobs involved simply shifting matter from one place to another. But today, 35 per cent of jobs involve deeper thinking.

Another set of changes over time involves the home, and physical health. Children now get better nutrition during their formative years when the brain is growing – for example via adding iodine to salt. Smaller families mean that parents can theoretically spend more time with and money on the fewer kids. A higher standard of living can mean fewer infections, so that children's potential growth is not hindered. A 2010 study showed a strong link between early childhood vaccinations and the average IQ of a nation.

OUR GRANDPARENTS ARE "STUPID"?

This Flynn Effect (of increasing scores on IQ tests) does seem to be real. It also seems to kick in strongly when countries get to a certain level of health, education and welfare.

But look at it another way.

IQ is rising at 3 points per decade, 9 points per generation (assuming 30 years for a generation), and so we can extrapolate to 33 points per century.

Does this mean that there is an ever-widening gulf between our minds and those of our parents and grandparents?

One century ago, what would our great-great-grandparents have scored, on average, if they took one of our modern IQ tests? Would they have IQs of around 70, or less?

Superficially, this implies that one century ago, most countries were beset by a plague of "mental retardation". How could those people, one century ago, follow a game of football or cricket if they were not intelligent enough to understand the rules?

This is obviously crazy. After all, they developed some of those games.

SO WHAT IS IQ?

One way to interpret this is to see your IQ not as something set in stone and unchangeable but, rather, more like a muscle that can alter itself and adapt to a changing environment. So IQ is a measure of how well we can deal with the current society we inhabit. In other words, perhaps it's a measure of how "modern" we are.

But we still don't understand the full ramifications. According to James Flynn, if these gains in IQ are actual and real, "Why aren't we undergoing a renaissance unparalleled in human history? . . . Why aren't we duplicating the golden days of Athens or the Italian Renaissance?" Perhaps we are, but we can't see it because we are in the middle of it?

In 2013, James Flynn described this increase in IQ scores in a TED talk. "The cars that people drove in 1900 have altered because the roads are better and because of technology . . . And our minds have altered, too. We've gone from people who confronted a concrete world and analysed that world primarily in terms of how much it would benefit them to people who confront a very complex world."

So okay, it's not as easy as ABC to test IQ . . .

And maybe teenagers were right all along, and they really do know more than their parents.

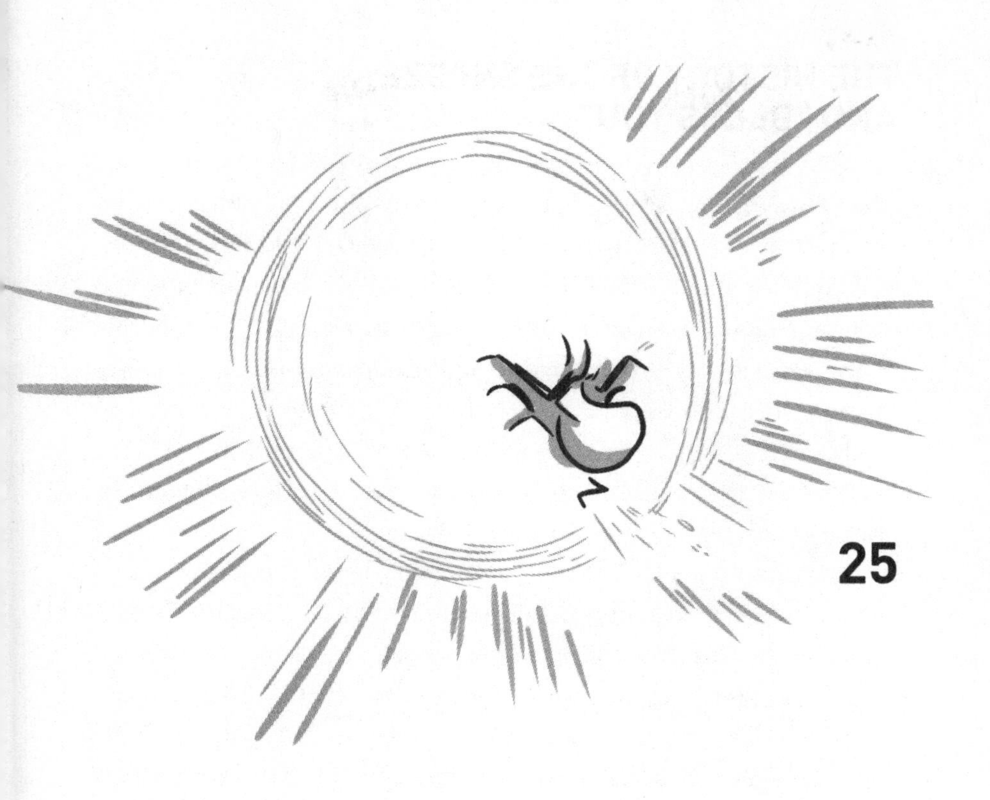

25

SNEEZE IN SUNLIGHT

If you do crosswords a lot, you'll know many unusual words. So cruciverbalists know that a fancy word for "sneezing" is "sternutation". But one thing that nobody knows is why some people sneeze when they go out from the dark into the sunlight. Welcome to the Photic Sneeze.

By the way, it also happens with artificial light, not just sunlight. Furthermore, the colour of the light doesn't matter – just the change in brightness from dark to light.

THE HISTORY OF THE SNEEZE AND "BLESS YOU"

They famous Greek physician Hippocrates thought that sneezing would cure hiccups. Soranus, another Greek physician, thought that sneezing could sometimes help treat epilepsy. Celsus of Rome thought that it was a sign of recovering from an illness.

In general, both the Greeks and the Romans looked upon sneezing as a sign of good health. So if a neighbour sneezed, you would express your best wishes to that person with the phrase "Live long" or "May Jupiter bless you". So that's one explanation of today's phrase of "Bless you" after somebody sneezes.

But while some ancient sources considered the sneeze both beneficial and therapeutic, others deemed it to be a bad omen, and associated it with malevolence and disease.

In that spirit of alternative interpretations, let me also give you a completely different origin for the phrase "Bless you". Apparently, some Pagan cultures believed that when you sneezed, your soul temporarily left your body. So anybody nearby had a duty to protect your body from being taken over by evil spirits by saying the magic words "Bless you".

And just to show how accommodating and open-minded I am, here is a third explanation.

In the 14th century, the plague known as the Black Death ravaged Europe. One of the Roman Catholic popes apparently declared that the faithful should say "May God bless you" immediately after a sneeze to protect the sneezer from the plague. Unfortunately, the short prayer was unsuccessful, and one third of all Europeans died. The real culprit was not the sneeze, it was the bacterium *Yersinia pestis* carried by fleas on rats. (Read more on the Origin of the Black Plague in my 27th book, *Science is Golden*.)

A Sneeze Can Pop Your Eyes From Your Head?

Do you shut your eyelids when you sneeze specifically to stop your eyeballs popping out? No.

During the formation of a sneeze, there is an increase in pressure in your chest. Some of this pressure gets transmitted into the skull. An even smaller amount passes through into the back of the eye socket.

In terms of your anatomy, the globe of the eye is held in position quite securely by various muscles and the optic nerve. The pressure from the sneeze is just too small to rupture these muscles and the optic nerve.

For some unknown reason, a few folk have trained themselves to be able to sneeze with their eyelids open. Some people with Bell's Palsy also sneeze with their eyelids open. Their eyeballs still stay inside their eye sockets when they sneeze.

The physiologists would say that shutting the eyelids is just part of the Sneeze Reflex. But maybe it's so you won't see how ridiculous you look when sneezing, if you happen to be in front of a mirror.

HISTORY OF THE SUNLIGHT SNEEZE

A very small percentage of people will sneeze as regular as clockwork whenever they go into the light – always the same number of times – so they are strongly "compelled" to sneeze.

Around 350 BC, the great Greek thinker Aristotle tried to understand why sunlight would sometimes cause sneezing. He thought it was the heat of the Sun doing it.

In Book XXXIII of his *Problemata*, in the section entitled "Problems Concerning the Nose", he wrote, "Why is it that one sneezes more after one has looked at the sun? Is it because the sun engenders heat and so causes movement, just as does tickling the nose with a feather? For both have the same effect; by setting up movement they cause heat and create breath more quickly for the moisture; and it is the escape of this breath which causes sneezing."

In 1635, the English philosopher and scientist Francis Bacon also tried to work it out: "Looking against the Sunne, doth induce Sneezing. The Cause is, not the Heating of the Nosthrils; for then the Holding up of the Nostrills against the Sunne, though one Winke, would doe it; But the Drawing downe of the Moisture of the Braine. For it will make the Eyes run with Water; And the Drawing of Moisture to the Eyes, doth draw it to the Nosthrills, by Motion of Consent; And so followeth Sneezing . . ."

On one hand, Bacon showed that the heat was not the cause. His eyes needed to be open for it to happen. On the other hand, he thought that tears (induced by the bright sunlight) caused the sneeze. Unfortunately for his explanation, sneezing happens long before the tears would have had time to be generated.

It took until 1954 before Dr J. Sedan first described it in the medical literature. He called it "Photosternutatory Reflex". In 1964, Dr H.C. Everett published the first major survey of sunlight-sneezing.

He found it to be present, to varying degrees, in about one quarter of Harvard medical students (not exactly a typical population). In 1984, Dr Stephen J. Peroutka noted that three generations of his family had sunlight-sneezing. He called for further research to try to work out the mechanism.

In 1993, Dr R.A. Breitenbach wrote about it as a potential hazard for combat pilots. What if the pilot was flying through closely spaced clouds (giving alternating dark and light stimuli to the eyes), or getting a sudden reflection of the sun from a body of water while landing? It could also be a hazard for other professions, such as high-wire trapeze artists.

Disease Transmission

Surprisingly, we still don't fully understand the basic mechanisms by which even common respiratory diseases get transmitted. There seem to be two main pathways.

On one hand, the people (infectious person and innocent bystander) need to be physically close. "Self-Inoculation" happens when their mucous membranes come in direct contact, or when bodily fluids are transferred. "Large Droplet Transmission" arises when infected droplets travel through the air and are sprayed "directly onto the conjunctiva or mucous [membrane] of a susceptible host via coughing or sneezing".

The other way, "Airborne Transmission", is both indirect and long range. For example, a sick person could sneeze while inside an elevator, and pass on their disease to others who later enter that elevator. The infected person coughs or sneezes small pathogen-carrying droplets into the air. These

droplets, after a time, are breathed in and enter the airways of the innocent bystander. Alternatively, the droplets can evaporate and shrink into pathogen-carrying solid residues called "droplet nuclei" (about 5 to 10 micrometres across). These can also infect another person.

TORTURED ACRONYM – ACHOO

One official names for sneeze-when-you-enter-the-light is the "Photic Sneeze Reflex". But there's another one.

In 1978, W.R. Collie whimsically came up with the "ACHOO Syndrome" while at a conference dinner. (You know how convenient it is to make notes on the paper napkins at conference dinners . . .) "ACHOO" is the sound you make when you sneeze.

So with a bit of imagination, this medico made the "A" in "ACHOO" stand for "Autosomal Dominant". Correct: the Photic Sneeze is genetically dominant. If one parent has it, then their children have a 50 per cent chance of also having it. There seem to be associations with genes on Chromosomes 2 and 15.

The "C" stands for "Compelling". It has the advantage of starting with the letter "C", but the reflex is not all that compelling. Roughly one quarter of us have this strange reflex of sneezing in the light – but there's a wide spectrum of responses.

A very small percentage of people will sneeze as regular as clockwork whenever they go into the light – always the same number of times – so they are strongly "compelled" to sneeze, whether they want to or not. At the other end of the spectrum are people who are not compelled; they have control over this reflex. Suppose they are just on the point of sneezing for some other reason, with the inside of their nose tickling away like crazy. They will sneeze only when they deliberately then stare into a very bright light. So "C" for "Compelling" is more artistic than accurate.

Next in the word "ACHOO" is the letter "H". This stands for Sun, from the Greek word for "Sun", "Helios". That's not too big a stretch.

A sneeze is a protective reflex that "cleans" the upper airways.

Finally, we have the two "O"s at the end of "ACHOO". The first "O" stands for "ophthalmic", coming from the Greek word "ophthalmos" meaning "eye". I guess the "eye" got in there because you have to see the sun with your eyes . . .

The second "O" stands for "outburst". Okay, an "outburst" could be loosely thought of as a "sneeze".

1045 Kilometre-per-hour Sneeze!

The maximum exit speed of the air from your mouth and nose is often stated to be around 160 kilometres per hour – with a snot delivery range of about 10 metres. (I have seen figures of 320 kilometres per hour in some literature, and even 1045 kilometres per hour in a peer-reviewed journal!) However, when the MythBusters (Jamie Hyneman and Adam Savage) tested sneezes, they could achieve only 65 kilometres per hour and a distance of 5 metres.

I went looking through the peer-reviewed literature. I checked journals in fields as varied as Respiratory Medicine, Thermal Engineering, Laser Engineering, Fluid Dynamics and Homeland Security. The highest trustworthy figures I could find were around 45 to 50 kilometres per hour.

SNEEZE PHYSIOLOGY

A sneeze is a protective reflex that "cleans" the upper airways.

This reflex is exquisitely coordinated, involving many nerves and muscles firing off at exactly the right time. The fact that it's a reflex means that you don't have a lot of control over it.

A sneeze has four parts.

First, the sensory part. The mucous lining of the airways (usually in the nose) is stimulated. The stimulation can come from an internal source – say, histamines or leukotrienes, released from mast cells or eosinophils (a type of white cell). The trigger can also be external sources – such as fine dust or irritants, smoke or odours, or cold air. These triggers land on your respiratory mucosa and stimulate branches of the Fifth Cranial Nerve (the Trigeminal Nerve). But the stimulus can be extremely varied. It can range from the plucking of your eyebrows, having an anaesthetic injected into your skin near the eyes, sexual excitement or orgasm, and even a big meal, right up to, yes, the classic stepping out into bright light.

The second part of a sneeze is the inspiratory part. Air is sucked deeply into your lungs, while your eyes automatically close.

In the third part of the Sneezing Process, you try to empty your lungs. But your vocal cords snap shut, and so the air pressure builds up inside your lungs.

In the fourth and final stage, the vocal cords open suddenly. The pressurised air bursts out of your nose and mouth in a turbulent blast. This scours the lining off the airways, and hopefully removes nasal debris and irritants.

THEORIES – PICK ONE

So what makes people sneeze in bright light? At this stage, we simply don't know.

One theory says that in some locations in your brain, the nerves that set off a sneeze are physically close to the nerves that carry visual information. The theory then says that information "leaks" between the separate nerves. This leaking is also called "cross-talk". So the nerves-that-carry-light-information accidentally fire off the nerves-that-stimulate-sneezing.

The other major theory says that it's related to one of your major Nervous Systems – the Parasympathetic Nervous System. We know that the Parasympathetic Nervous System makes your pupils smaller when light lands on your eyes. It is also involved in triggering sneezes. Again, various locations have been blamed for cross-talk from one nerve to another.

But overall, we still don't know why some of us sneeze in the Sun.

Maybe it's a evolutionary hangover from when we were cave-dwellers, living with a smoking fire. A good sneeze would have felt wonderful. Indeed, a few people with the Photic Reflex have told me that they deliberately start the day by looking briefly at the Sun. They get "cleaned out in the airways" – and get a "feel good" moment from the light.

We do know that Photic Sneezing can be hazardous. Imagine that you are a strong Photic Sneezer driving a car coming out of a tunnel, or through the alternating light-and-dark generated by a line of trees at the roadside. In this case, you could wear a hat or polarising sunglasses to shield your eyes.

Would Law Enforcement officers accept the Photic Reflex as a valid excuse for driving into somebody's car? Would saying "I'm sorry, Officer, but I have the ACHOO Syndrome" actually cut it?

Yes, if your local court will accept the somewhat unusual legal defence of "non-insane automatism". The UK case of *R v Woolley* dealt with a truck driver's sneezing fit causing a seven-car pile-up. The driver was acquitted, because his sneezing was out of his control.

No, if your local court reckons that it's not reasonable to continue driving if you're having a sneezing fit.

ACHOO-ers Are Different

It seems that Photic Sneezes have a lower threshold to be visually excited. Research shows that when Photic Sneezers see the world around them, their visual cortex responds in a more excited manner than non-Photic Sneezers.

So maybe they need both the slightly more excitable brain, as well as the cross-talk between the "light nerves" and the "vision nerves"?

SPACE, THE HOSTILE FRONTIER (MORE THAN A BAD HAIR DAY)

The TV series *Star Trek* referred to space as the "Final Frontier". *Star Trek* was set far in the future.

By 2014, just over 500 humans had gone into the microgravity of space. They have spent a total of only about 120 person-years up there. Yet that was enough to teach us that *Star Trek* was wrong in its naming, and that space should really be called the "Hostile Frontier".

The environment inside the International Space Station (ISS) as it zips around the Earth once every hour and a half seems perfectly benign. After all, the astronauts get around wearing casual clothing, such as shorts and T-shirts.

But underneath the clothing, strange things happen to the human body in space.

Microgravity, not Zero Gravity

The astronauts on the ISS are *not* in zero gravity. They are actually very close to the Earth, and well inside its gravitational field. You and I are about 6370 kilometres from the centre of the Earth. The astronauts are only another 400 kilometres further out – just 6770 kilometres from the centre. So they experience about 90 per cent of the gravity they would on the surface.

So how come they "float"?

Because they are always falling to Earth, and always missing it!

On one hand, in each second they "fall" 4.5 metres.

On the other hand, they are travelling at 27,700 kilometres per hour "parallel" to the ground. That's about 7 kilometres per second. In each second, they travel 7 kilometres across the curved surface of the Earth. Over that 7 kilometres, the curve of the Earth "drops" by 4.5 metres.

So the astronauts (and the ISS) are always falling 4.5 metres in each second, but the Earth is "curving" away from them by the exact same amount.

(A full explanation of this is in the story called "Zero Gravity" in my 28th book, *Never Mind the Bullocks.*)

By the way, the gravity on the Moon is 16.5 per cent of Earth's gravity. On Mars, it's about 38 per cent.

BONES IN MICROGRAVITY

Consider the astronaut Richard Hieb. At 6'3" tall, he was just one inch under the official NASA height limit for astronauts of 6'4". (The USA uses Imperial Units of Measurement, not Metric.) On 8 July 1994 he went into orbit on the Space Shuttle *Columbia*. Within four days he had grown more than an inch in height – and so was "technically" too tall to fly on the Space Shuttle.

You see, while he was walking around on Earth, gravity compressed the discs in his spine. But in the microgravity of space, that compression was removed. Astronauts in space have grown an extra 2 inches (or 50 mm) in height – on average, roughly 3 per cent taller. That extra height can stretch the sciatic nerve and cause excruciating back pain.

Hieb Talks to Ground Control

The following conversation was reported by the *New York Times*:

"According to my quick calculations here, I seem to have grown about an inch or so. So I'm now too tall to fly in space," Mr Hieb informed payload controllers after measuring himself. "And that's without slipper-socks."

A ground controller in Alabama said, "I just hope the flight director's not listening."

"*We heard that*," answered a voice from Mission Control in Houston.

Bone is living, dynamic tissue. Down on Earth, the mere act of standing or walking puts stresses on our bones. Bone responds by rebuilding itself to oppose those stresses. But in the microgravity of space, you lose those continual stresses. As a result, bone both rebuilds itself to be less dense and also loses minerals.

Bone Turnover

The human body has two major types of bone. They destroy and rebuild themselves at very different rates.

"Cortical" bone is very compact and dense. It varies between 70 and 95 per cent solid (or 30 and 5 per cent porous). It makes up the shaft of long bones – smooth, white and solid. More specifically, it forms the outer shell around the bone marrow. It makes up 80 per cent of your skeletal mass. Down here on Earth, cortical bone has a very low turnover rate of just 3 per cent. That means that 3 per cent is destroyed, and rebuilt, each year.

"Trabecular" bone is in the bone marrow cavity (which exists in ribs, vertebrae and the ends of long bones). It has a very open and porous structure – varying between 10 and 70 per cent solid. That leaves lots of room for the blood vessels and the marrow. Trabecular bone makes up only 20 per cent of the mass of your skeleton but, thanks to its porous structure, has 10 times more surface area than cortical bone. If the external stress on trabecular bone changes, it adjusts by rearranging its internal structure. It has a Bone Turnover Rate of 25 per cent per year.

Two things happen to bone in microgravity.

First, bones weaken enormously, and very quickly. For example, down here on Earth, postmenopausal women can lose 1 to 1.5 per cent of their hip bone mass in one year via osteoporosis. But an astronaut in microgravity can lose 1 to 2 per cent in a single month, in a process similar to osteoporosis.

Astronauts are at risk of fractures upon their return to Earth. (Some *Mir* astronauts on extended missions lost 20 per cent of their bone mass.) Luckily, when they return to Earth, they can recover most of their bone mass. But the bone is laid down in a more porous fashion. It will never be as strong as before their space journey. They will have a slightly greater risk of bone fractures for the rest of their lives.

The second problem with bone loss in space is that the minerals astronauts lose from their bones have to go somewhere. Often, they can find their way to the kidneys. Once there they can cause the very painful condition of kidney stones. Kidney stones can be so debilitating that they could end a space mission. Between 2001 and 2014, 14 astronauts on space missions developed kidney stones. Luckily, none were severe enough to force an emergency return to Earth.

ABSENCE (OF GRAVITY) MAKES THE HEART GROW ROUNDER

Microgravity changes astronauts' fluid circulation – and the shape and strength of the heart.

Here on the Earth's surface, gravity pulls the various fluids in our body downward. But in the microgravity of space, the fluids rise – so astronauts' legs shrink, but their faces swell. The extra fluid can cause headaches.

Fluids also accumulate in the nose. Astronauts say the sensation is like hanging upside down back on Earth. Perhaps this nasal congestion is why astronauts report that their sense of smell is weakened, and why food tastes bland.

> Astronauts in space have grown an extra 2 inches (or 50 mm) in height – on average, roughly 3 per cent taller.

On a more serious note, the heart shrinks and changes shape – it becomes more spherical and functions less efficiently. Astronauts can lose one quarter of their aerobic capability after just a two-week mission in orbit. The total blood volume (and number of red blood cells) can reduce by as much as 20 per cent. Some astronauts have also suffered abnormalities in their heart rhythms during spaceflight.

Others have fainted up to several days after they returned to Earth, because their blood pressure fell so low. If an astronaut fainted during re-entry from orbit, while they were hand-flying the spaceship, it would be disastrous.

EYE CHANGES

For some astronauts, microgravity changes their vision – permanently.

In October 2009, astronaut Mike Barratt returned to Earth. His eyesight had changed profoundly. Previously, he needed glasses for distance vision, but not for reading. But after six months on the ISS, his vision had reversed. He was now eagle-eyed for distance vision, but needed glasses for reading.

The cause? Remember how in astronauts the bodily fluids shift upward from the legs. The increased pressure inside Barratt's skull forced the backs of his eyeballs to become less round and more flat. As a result, when he's reading, the image processed by his eye now comes to a focus behind his retina, giving him blurry vision.

Some 60 per cent of astronauts have eye damage as a result of microgravity. About a dozen, like Barratt, have permanent changes. Each of Barratt's retinas now has microscopic folds or wrinkles. His eyeballs aren't round anymore, but flattened at the back.

We don't know why this change affects some, but not all, astronauts. We also don't know why (so far) it affects only male astronauts, not female astronauts.

MUSCLE

I love muscle. It's so amazing.

Consider this. No matter how much you use an old, slow car, it will never turn into a sports car. But the more you use muscle, the stronger it gets.

Some 60 per cent of astronauts have eye damage as a result of microgravity.

In microgravity, muscle loss is a major problem. Like the old cliché, if you don't use it, you lose it.

Muscles are made up of lots of muscle fibres. Down here on Earth, the activities of daily living keep your muscle fibres in good condition. But even short spaceflights of less than 14 days are enough to make muscle fibres shrink by one-third. Longer missions can shrink muscle fibres by as much as 40 per cent.

And this happens even though the astronauts perform specific exercises aimed at keeping their muscles strong.

One study looked at astronauts who were in microgravity for six months. The astronauts did both aerobic exercise (five hours per week) and specific resistance exercises (3 to 6 days per week). Their allocated exercise time was up to 2.5 hours per day.

Even so, averaged out over all the astronauts, the peak power of their calf muscles had dropped 32 per cent by the end of their stay.

RADIATION

We are not normally aware of this, but the Earth's atmosphere and magnetic field shield us from radiation. In space, you are no longer protected by the Earth's atmosphere or magnetic field. So you get exposed to lots of ionising radiation.

There are many different types of ionising radiation in space. At this stage we know of three relevant types that can affect astronauts.

The dominant type for astronauts on the ISS is Galactic Cosmic Radiation (also called Cosmic Rays). GCR is penetrating particles coming from Outer Space at nearly the speed of light. They are high-energy protons (85 per cent), helium (14 per cent), and heavy nuclei and ions (1 per cent). The Sun cycles into peak activity every 11 years or so. This is called the "Solar Maximum". At Solar Maximum, the Sun's magnetic field tends to protect us more from the GCR.

The next type of radiation is "Solar Particle Events". SPEs are mostly medium-energy protons shot out from the Sun during solar flares.

A Russian cosmonaut, Sergei Avdeyev, was terrified by radiation from SPEs while orbiting in the Russian space station *Mir*. (He spent a total of 747 days in space. On one trip, he was up there for 379 days.) He said, "I felt that the particles of radiation were walking through my eyes, floating through my brain, and maybe clashing with my nerves."

Finally, there are charged particles trapped in the Van Allen Belts. The Van Allen Belts look like two nested doughnuts, one inside the other, surrounding the Earth. They reach from an altitude of 1000 kilometres to 60,000 kilometres above the Earth's surface. The inner one contains mostly energetic protons. The outer one contains both protons and electrons. They are a source of radiation – but only if you travel through them. The ISS, at only 400 kilometres above the surface, is usually not affected by the Van Allen Belts.

The ISS was designed with in-built shielding to protect the astronauts in the event of a sudden increase in radiation. They would retreat to the storerooms in the centre of the ISS and huddle behind internal metal walls, containers of food, etc. This kind of shielding is not effective against GCR because it is so powerful. But it can protect against SPEs.

There are different ways of measuring radiation, but the most useful is the milli-Sievert (mSv). The average Australian gets a natural background radiation dose of about 1.5 mSv per year – equivalent to about five chest X-rays. But in space astronauts can be exposed to much higher levels. Over a six-month stay they could absorb 80 mSv (at Solar Maximum) or 160 mSv (at Solar Minimum).

A 100 mSv dose spread over a few minutes will give you a 1-in-200 chance of developing a cancer later in life. A single dose of 1000 mSv lifts the cancer risk to 1 in 20. An acute dose of 3000 to 5000 mSv will kill half the people who receive it over a few months.

A 10,000 mSv dose will cause death in a few weeks.

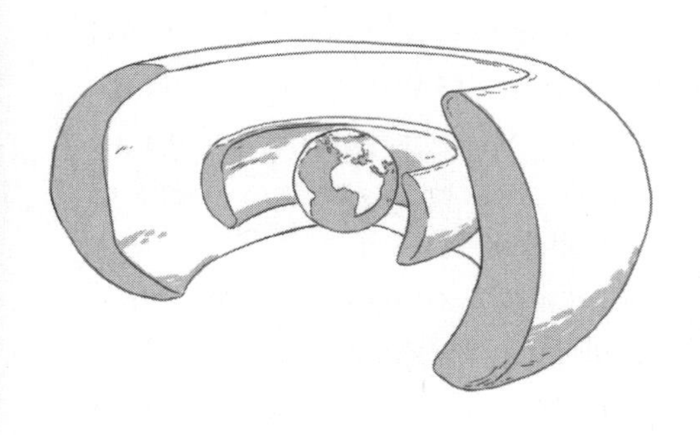

The Van Allen Belts look like two nested doughnuts, one inside the other, surrounding the Earth. They reach from an altitude of 1000 kilometres to 60,000 kilometres above the Earth's surface.

Deadly Van Allen Belts

The We-Never-Went-to-the-Moon Denialists wrongly claim that the Moon astronauts would have died from radiation poisoning as they traversed the Van Allen Belts. The astronauts would have crossed them twice – on the way there, and on the way back.

Yes, a radiation dose of 3000 mSv, delivered in one hour, will kill half of the recipients. But the radiation dose delivered to the outside of their spaceship from the Van Allen Belts was much lower to 130 mSv. Furthermore, the Moon astronauts were protected by the metal walls of their spacecraft, and so received a dose of only 20 mSv. It is a moderate dose of radiation – but definitely much too low to kill you in the short term.

BALANCE

Some astronauts get Motion Sickness once they enter microgravity. In some cases, the cause is that Otoliths (tiny "bones" in the Balance Organs that respond to motion) respond differently to movement in microgravity.

> The Earth's atmosphere and magnetic field shield us from radiation. In space, you are no longer protected.

Some astronauts, after their return to Earth, become faint or dizzy when they stand upright. This usually lasts for some time. The cause is that their Central Nervous System cannot properly control how and when arteries constrict. This leads to abnormal regulation of blood pressure, which leads to the brain being starved of blood, which leads to dizziness.

IMMUNE SYSTEM

The Immune System changes in space. Unfortunately, so do the bad guys it has to fight.

Unluckily, the bad guys become nastier, and harder to defeat.

In microgravity, nasty bacteria (*Salmonella*, *Pseudomas*, etc.) evolve to grow faster, and to be more virulent. They also form themselves into strange multi-layer biofilms of a type never seen on Earth. This new structure gives them resistance to the chemicals and cells that our Immune System manufactures to deal with invaders.

Then the functioning of the Immune System is also altered in space – and for the worse. For example, many of us carry herpes, which hides out inside our Central Nervous System. Our Immune System normally keeps herpes at bay. But something changes in microgravity. Astronauts often have their latent herpes viruses reactivated – even during short-duration spaceflight.

Luckily, within two weeks of landing, Immune System responses seem to return to normal.

THE FUTURE

I believe that we humans will eventually become a space-going race. At some stage more humans will be living off the planet than on it.

> Astronauts often have their latent herpes viruses reactivated – even during short-duration spaceflight.

After the ISS, there will be missions to the Moon (again) and then to Mars. Commercial space flight is rapidly increasing, as measured by the number of space flights each year. Looking further afield and further into the future, there are good arguments for mining asteroids.

Earth has been our cradle. We are slowly learning how to grow up and leave home – and growing pains are a normal part of that.

Earth – Our Home

Let me be very clear.

On one hand, I firmly believe we will become a space-roving race. This will happen once we can access more energy. For example, there is more energy in one cubic metre of anything (due to the Higgs Field) than the Sun emits in 1000 years. At the moment, we have no idea how to get at this energy. Being able to access huge amounts of energy will be the key to space travel.

On the other hand, future space travel is not an excuse to mistreat our planet. The Big Danger right now is Global Warming. We should leave our home in better condition than we found it.

27

SWEARING AND PAIN

WARNING!
THIS STORY MAY CONTAIN TRACES OF SWEARING.

Swear words, or bad language, make up about 0.6 per cent of our spoken language. Given that we speak an average of about 16,000 words each day, this means that about 95 of our daily words are profanities. In general, swear words are intended to be offensive – but there is one specific situation where they are actually very helpful.

Where Do Swear Words Come From?

In our English language, the majority of swear words we use come to us from German language roots, not Latin.

BAD WORDS 101

Perhaps we swear precisely because we are told not to?

The word "profane" comes from the Latin roots "pro" meaning "before" and "fanum" meaning "temple". So a profanity was something that you said only "before" the temple – or, in today's language, "outside" it. A profanity was definitely not to be spoken inside the temple.

Swear words do have power. Merely hearing profanities will change the electrical conductance of your skin.

Cursing is universal. Profanities exist in every single language ever studied. Every language, dialect or patois has "forbidden" or bad words – regardless of whether that language is living or dead, and whether it is spoken by billions or just a small tribe.

Interestingly, some cultures draw their swear words from religion. For example, when four centuries ago Shakespeare wrote "zounds" or "'sblood", he was using very offensive (for the day) contractions of the phrases "God's wounds" for "zounds", and "God's blood" for "'sblood".

Other societies fiercely protect the concept of the honour and "purity" of women. So, many of their swear words relate either to female genitalia, or to the theme of "son of a whore".

POWER WORDS

Swear words do have power. Merely hearing profanities will change the electrical conductance of your skin. Your pulse will quicken, the hairs on your arms will rise and your breathing will become shallow.

But languages evolve. So the power of swear words can change over time.

Nobody today would be bothered by the word "golly". However, originally, that word was a very obscene and profane contraction of the phrase "God's body". So that swear word has weakened.

Sometimes, it goes the other way. Neutral words can become a little unpleasant or uncomfortable to use. For example, the word "coffin" originally meant a "box". But once the word "coffin" became linked to the concept of death, people stopped saying "Let's think outside the coffin", or "Let's see if there's anything to eat in the bread coffin". I think that's a gosh darn shame.

SET UP THE STUDY

Back in 2009, Dr Richard Stephens and colleagues from Keele University in the UK looked at the link between swearing and pain. Lots of people swear when they suffer an injury and suddenly feel pain.

This was the scientists' question: does the swearing relieve the pain, make the pain worse, or have no effect on the pain at all?

They got 67 unfortunate undergraduate students to undergo a standard pain test called the Cold Pressor test. In the Cold Pressor test, the students were instructed to submerge their unclenched and non-dominant hand into cold water, for as long as they could stand. How cold? Bloody cold: 5°C.

And while their hand was in the cold water, the students were to say, over and over, either a swear word or a neutral (Control) word.

How did Stephens and Co. pick the swear word? The students were each asked for "five words you might use after hitting yourself on the thumb with a hammer". The experimenters chose the very first swear word on their list to be that person's naughty word.

Why have a Control word? The experimenters needed to control for the possibility that simply saying any word at all could change how long the subjects would be able to keep their hand in freezing water. So they also asked the students for "five words to describe a table", for example, bench, counter, desk, worktable, horizontal surface and so on. They then chose one of these words for each student to use as a Control word.

Further Study?

So far, only a few studies have looked at the relationship between saying certain words and whether the feeling of pain is changed.

They have definite limitations.

First, their subjects are always the easy-to-get ones – junior year psychology students. I would love to see a study involving tradespeople who get physical injuries on a regular basis as a result of their work. Second, the sample sizes are small. Third, there seems to be a wide scatter in the results. In one case, the standard deviation was equal to the average! (If you know your statistics, you would know that this is a little "dodgy" or unreliable.)

Let's see what further study reveals.

RESULTS – TIME AND HEART RATE

Stephen's and Co.'s study showed both the presence and absence of a distinct gender difference, with regard to time.

> If you're prone to cuss and curse, you would get less pain relief by swearing.

First, while repeating over and over a word meaning "table", on average, men could withstand the cold water for a bit over two minutes (146 seconds). However, women could go only for a bit over one minute (83 seconds).

Second, if they repeated their chosen profanity, each gender could keep their hand in the 5°C water for an average of an extra 40 seconds. Swearing gave both the males and females approximately the same increase in time in the cold water.

A similar pattern was found for the heart rate.

There was a difference. Comparing men and women, in each case the men had a lower heart rate before the Cold Pressor test, with the Cold Pressor test and the Control word, and with the Cold Pressor test and the swear word.

But there was also a similarity. For both sexes, their pulse was lowest before the Cold Pressor test. It rose a bit when they had their hand in cold water and were saying a neutral word, but rose again even higher when they were saying their preferred profanity. The increase in the heart rate was approximately the same for each sex. This was probably related to the release of adrenaline. However, no blood samples were taken to test these levels.

So, according to this study, swearing makes pain "go away" or become more "bearable" – because the subjects could keep their hand in cold water for longer.

Good Words or Bad Words?

It seems that perhaps shouting out Bad Words can relieve pain. What about the opposite – Good Words?

A 2005 study explored (among other things) the effects of spiritual meditation upon pain. This study had some limitations, but it seemed to show that Good Words could relieve the pain of the Cold Pressor test. The subjects could choose one of four phrases, for example "God is peace", "God is joy", "God is good", and "God is love". One subject chose to substitute "Mother Earth" for "God".

Sure enough, good words also seemed to relieve pain – at least in this limited study.

But this leaves us with a mystery. Why do Bad Words and Good Words work to relieve pain – but not Control words?

THE POWER TO CHANGE PAIN

For better or for worse, swear words are more common nowadays than in the recent past. After all, we use swear words to express surprise and happiness, or anger and disgust.

But what about people who tend to swear more than average? Does the power of profanity fade the more you use it?

In 2011, Dr Stephens did a follow-up to see "if overuse of swearing in everyday situations lessens its effectiveness as a short term intervention to reduce pain". The answer was yes. The more the subjects swore on a daily basis, the less extra time they could withstand cold water while swearing.

So if you're prone to cuss and curse, you would get less pain relief by swearing.

Here is the lesson for the day: it's okay to profane, but only when you're in pain.

Ultimate Recognition

In 2010, Drs Stephens, Atkins and Kingston were awarded one of the highest honours known to Science. They were called to Harvard to receive the 2010 Ig Nobel Peace Prize. The Ig Nobel prize is awarded for research that cannot or should not be replicated.

Not to blow my own horn, but in 2002, I also received an Ig Nobel Prize – in the field of Interdisciplinary Research. It was for my ground-breaking research: "Belly Button Fluff – What Causes It, and Why Is It Almost Always Blue".

Harvard University showed me such great respect that they offered to fly me to Harvard at my own expense. They would not insult me by offering to pay for my flight.

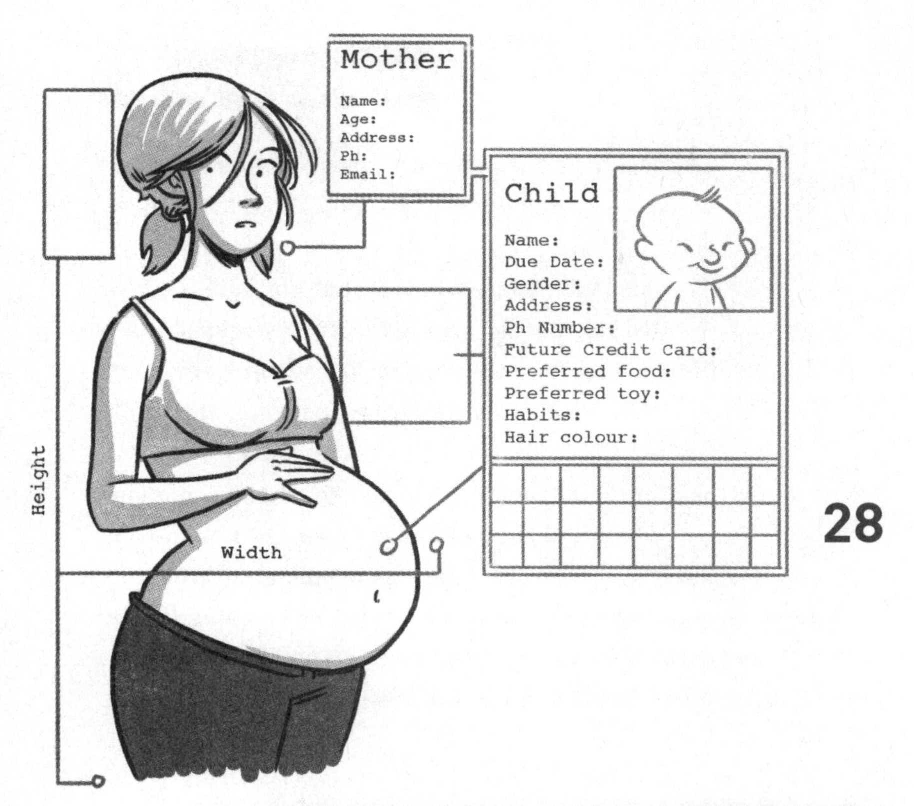

28

TARGET KNOWS BEFORE IT SHOWS

Whenever we do anything in our brave new digital world, we leave behind a rich and deeply textured trail of data. Our mobile phones continually tell the phone company where we are. (And, yes, all our phone conversations and texts are digitally recorded – just in case.) Our credit cards reveal how much we spend, and where, and on what. Our web browsers divulge all the sites we have visited.

Welcome to the world of Big Data, where the chain store Target has data-mined its way into my favourite organ, the uterus – and can now predict when you're having a baby!

Future Predictions

An old Danish proverb says, "It's very difficult to make predictions, especially about the future."

This is the basic problem that Big Data is trying to solve – predicting the future. It works at many different levels – military, traffic control, espionage, medical, retail sales, policing, and so on.

On the other hand, it's much easier to make predictions about the past – after the event has happened. Consider the Nostradamus-type prediction: "When the blue ducks land on the pond at sunset, the red ripples spread in all directions." To some people, it obviously predicts the death of Diana, Princess of Wales in a tunnel in Paris.

LITTLE DATA

Back in the old days, a bookstore might slowly build up a base of loyal customers by offering discounts, and mailing out a brochure every month or so. It worked, but it was no Amazon.com.

Little Data was good enough to predict the final quality (and price) of Bordeaux wines. These wines are initially astringent and acidic, and usually take years before they become drinkable. They are first aged in casks for 18 to 24 months, and then aged in bottles.

The problem was that it was very difficult to tell, immediately after harvesting, which vintages would blossom into greatness, and which would fall by the wayside. If the vintners could predict the price, they could buy low and sell high. But nobody could do this reliably, even when they factored in the taste of the wine from the cask and the weather in the harvest year.

This changed in the 1980s, when Orley Ashenfelter derived a "simple" formula. He was both a wine enthusiast and a Professor of Economics at Princeton.

$$\text{Wine Quality} = 12.145 + (0.00117 \times \text{winter rainfall (°C)} + (0.0614 \times \text{average growing season temperature (°C)}) - (0.00386 \times \text{harvest rainfall (mm)})$$

His formula, using Little Data, turned out to be much more accurate than actually tasting the wine from the cask.

BIG DATA INPUTS

Big Data is very different. Big Data involves data-driven statistical analysis and large-scale machine learning – both focused onto absolutely enormous amounts of data.

So let's look at the three Vs that underlie Big Data – Volume, Velocity and Variety.

First, Volume. In the USA, Walmart, the giant chain store, harvests some 2.5 PB of data each hour – just from customer transactions. A PB is a petabyte, and is equal to a million gigabytes. You could also think of a petabyte as 20 million filing cabinets full of files.

Second, Velocity. Today, we can analyse the data in real time. For example, you can predict the sales figures of big shopping centres by monitoring the activity of mobile phones in the car park – even before the customers have opened their wallets. You can also send text messages advising them of specials as they walk into the mall.

And third, Variety. The data can be GPS locations from your mobile phone, or it can be the entire contents of your address book with the phone numbers and birthdays of all your friends and their families. Big Data also includes everything you have ever

Three Vs underlie Big Data – Volume, Velocity, and Variety.

done surfing the Web or on email, Facebook, Twitter and all the other social media – and let's not forget the good old-fashioned credit card.

Your electricity supplier knows when you come home at night, when you cook dinner – and if you are growing marijuana under lights. Tracking companies already photograph your car's number plate so they can catalogue your movements. Cameras in shops and ATMs capture photos of random passers-by.

Improve Economy?

In their 2011 book *Race Against the Machine*, Erik Brynjolfsson and Andrew McAfee reviewed the previous decade in the USA. They noted that unemployment had increased, median wages had dropped, and that many people who lost their jobs could not find work that paid as much as their previous employment. These facts imply that the economists are not running the economies of their various countries succesfully.

I wonder what Big Data will predict for the economy up to 2021?

Could clever use of Big Data stop the recurrent Boom-and-Bust cycles in world economies? Are we smart enough to do that yet? I don't think so.

TARGET WANTS PREGNANT WOMEN

Back in 2002, Target in the USA wanted to know which of its shoppers had just become pregnant. Newly pregnant women (keen to buy New Baby stuff) are the Holy Grail for retailers.

Target (of course) wanted more customers. The difficulty is that shoppers tend to shop at the same place that they've always gone to – and it's hard to make them shift. But when women get pregnant, it's a whole new world – and a pregnant woman might be lured to buy at a different shop.

So Target hired Andrew Pole (who has Masters Degrees in both Statistics and Economics) to see if he could data-mine his way into freshly pregnant uteruses. His starting point was the so-called Guest ID number. This is a number that Target assigns to each customer, and is linked to their name, email address, credit card, age, marital status, house address, rental history, website visitation and more.

Andrew Pole, Statistician for Target

"We knew that if we could identify them in their second trimester, there's a good chance we could capture them for years. As soon as we get them buying diapers from us, they're going to start buying everything else too."

But that data is just the tip of the iceberg. According to a US Senate report, "private companies already collect, mine and sell as many as 75,000 data points on each consumer".

So Target (like anybody who will pay for it) can easily buy data about your employment history, race, preferred magazine and newspaper reading, education, car ownership and favourite online discussion topics. They can also buy data on your sexual orientation from clues in your posts on social media, as well as your political preferences, charitable giving, preferences in food and drink – and much, much more.

TARGET LOOKS FOR PREGNANT WOMEN

Big Data sitting on a hard drive is meaningless. Which is why Target employed Andrew Pole (and dozens of other statisticians) in their Guest Marketing Analytics Department.

Target had already set up Target Baby Registry. It soon became clear that women on this Data Bank were buying some 25 specific products around the beginning of their second trimester. These included mineral supplements such as calcium, magnesium and zinc, large quantities of unscented lotions, and so on. And if they bought a little bright blue rug, the new baby was probably a boy.

When they started buying cocoa-butter lotion, scent-free soap, hand sanitiser, washcloths, a handbag large enough to carry a few nappies and extra-large bags of cotton balls, they were obviously getting close to their delivery date. Indeed, Target found it could predict their delivery date with about 87 per cent accuracy – purely from their shopping habits.

So Target began sending discount coupons for various baby items to customers, according to how they rated on Target's own "Pregnancy Prediction Algorithm".

But then it got messy.

TARGET KNOWS BEFORE IT SHOWS

An angry father walked into a Target store just outside of Minneapolis, Minnesota clutching a brochure that had just come through the mail and demanding to speak to the manager. Furiously he said, "My daughter got this in the mail! She's still in high school, and you're sending her coupons for baby clothes and cribs? Are you trying to encourage her to get pregnant?" Sure enough, the brochure did carry the daughter's name and address, and had advertisements only for maternity clothing, nursery furniture, and so on. The manager apologised.

They can buy data on your sexual orientation, political preferences, reading habits, charitable giving, preferences in foods and drinks – and much, much more.

A few days later the manager called the father to apologise again. But this time, the father was a little embarrassed. He said, "It turns out there's been some activities in my house I haven't been completely aware of. She's due in August. I owe you an apology."

NSA Overwhelmed by Data

The American National Security Agency (NSA) is charged with all US electronic spying – outside the USA. This covers everything from listening to terrorist chatter to snooping on the communications of foreign governments. Apparently, the NSA is not allowed to snoop inside the USA.

In early 2014, the internet carried 1826 petabytes of information each day. According to the *Australian Financial Review*, the NSA "touches about 1.6 per cent of that. Of that amount, 0.025 per cent is selected for review – so NSA analysts look at 0.00004 per cent of the world's internet traffic."

That works out to about 1 terabyte per day.

NSA Snoops on Friends

In 2013, it was revealed that the NSA had been spying on foreign leaders, including the German Chancellor, Angela Merkel. She responded on 25 October, "We need trust among allies and partners. Such trust now has to be built anew."

Joe Trippi, a long-time American Democratic campaign worker and Fox News consultant, said on *Fox News* on 28 October, "The real miracle here, or stunning thing to me, is that Angela Merkel thought that she could talk on a cellphone and no one would be listening to her, allies or foes."

What useful information about the Enemies of the USA could be gathered by spying on the USA's friends?

TARGET PRETENDS IT DOESN'T KNOW

Now, some people like their privacy. (How quaint and 20th-century!) They get upset if they know they are being spied upon.

To be specific, Target customers were getting annoyed by receiving brochures advertising only baby merchandise, especially when they had not told anyone they were pregnant. So Target produced new brochures that had non-baby merchandise (wine glasses, lawnmowers, etc.) sprinkled around the baby goodies. As a result, the newly pregnant mothers didn't realise they had been data-mined.

Target is happy. Their revenue jumped from $44 billion in 2002 to $67 billion in 2010. This was helped mightily by the fact that with regard to pregnancy, Target knows before it shows.

As Andrew Pole said, "Just wait. We'll be sending you coupons for things you want before you even know you want them."

WHAT ELSE CAN DATA-MINING DO?

Forecast house prices? Openly available web search data on house prices can predict changes in metropolitan areas more quickly than real estate agents can.

Examine phone records to detect "terrorist" activity? Maybe.

Check airline statistics to work out the best time to buy tickets? Yes, and it's pretty close to accurate.

Snoop on what customers are searching for online to make useful predictions about their future behaviour? Yes, if their "behaviour" includes opening weekend box-office revenue for feature films, ranking of songs on the Billboard Hot 100, and video game sales in their first month.

Guess the Oscar winners? Sure.

Big Data can also help find people who are vulnerable as soon as a big storm hits, and help them. This would include those who need continuous electricity for their medical assistance devices, those who have certain types of disability, and so on.

PHILOSOPHICAL MEANDERINGS

One Big Problem with Big Data is that Correlation does not mean Causation. In other words, just because two factors look related does *not* mean one caused the other.

Big Data can also help find people who are vulnerable as soon as a big storm hits, and help them.

For example, in the USA for the years 2006 to 2011, there's an excellent statistical correlation between the dropping murder rate and the declining use of the browser Internet Explorer.

Furthermore, between 2000 and 2009, there's another magnificent correlation between the dropping divorce rate in Maine and US-wide per capita rates of consumption of margarine.

Are these factors correlated? Beautifully.

Did one cause the other? Absolutely not.

So once you have a correlation, you need to look more deeply to see if A *caused* B.

One "advantage" of Big Data is that it takes you away from the dominance of "HiPPO" – the highest-paid person's opinion.

But a disadvantage is that Big Data removes our anonymity and privacy. If you use Google for email, Google automatically "machine-reads" everything you've typed. If you write about mattresses, advertisements for mattresses will usefully/annoyingly pop up. Even so, people still use Google's Gmail.

It's still early days for Big Data. But one important future step might be to increase transparency and openness about what we (the customer) are giving away – and what we get back in return.

And who will watch the watchers?

Big Data Finds Flu

Influenza kills some 250,000 to 500,000 people each year, worldwide. Each year in the USA, it kills some 3000 to 5000 people.

So the Center for Disease Control (CDC) continually monitors the level of Influenza-Like Illness (ILI) in the US population. Some 2900 healthcare providers are registered participants in the CDC ILI Surveillance Program. The problem is that their information might be too late to be useful – the ILI warnings have a lag-time of 7 to 14 days.

In 2008, Google Flu Trends (GFT) was launched. It tried to link random enquiries on Google's search engine about the flu and its symptoms to actual CDC ILI statistics – but without the delay. Initially it was quite accurate, but for various reasons faltered both in 2009 and in the 2012/2013 epidemic.

In 2014, a more accurate set of predictions was made by analysing searches for "flu" and its symptoms on Wikipedia.

29

THE WORLD'S MOST EXPENSIVE BOOK

You might remember that Sarah Palin, the former governor of Alaska, was nominated as the Republican Vice-Presidential Candidate for the 2008 US Presidential Election. But she never did get to be "one heartbeat away" from being President of the United States, because Barack Obama and the Democrats won the 2008 Election.

I vividly remember one comment Sarah Palin made about scientific research. She said, "Sometimes these dollars go to projects that have little or nothing to do with the public good, things like fruit fly research in Paris, France. I kid you not." Surprisingly for an advocate of Free and Unregulated Markets, she didn't foresee how much value was generated by research into fruit flies. (Clearly, this was the biggest problem with Sarah Palin's campaign.)

Within a few years, a book on fruit flies would become (temporarily) the world's most expensive book.

FRUIT FLY BIOLOGY 101

Like most of us, Sarah Palin probably didn't realise that the common fruit fly (*Drosophilia melanogaster*) is one of the most widely used organisms in genetic research. In fact, the fruit fly is incredibly important to scientists working in the field of Developmental Biology.

The Australian Fruit Fly is *Not* the Common Fruit Fly

It turns out that there are two families of "fruit fly" – *Drosophilia* and *Tephritidae. Drosophilia* is the one used as a model organism in genetic research.

***Drosophilia* is quite different from what Australians and South Africans call a "fruit fly". In these countries, the locals complain about flies belonging to a related but different family, *Tephritidae*.**

Back in 1910, Thomas Hunt Morgan began studying *D. melanogaster* in his soon-to-be-famous Fly Room at Columbia University. There were many reasons for choosing this particular creature. The common fruit fly has only four pairs of chromosomes (so reducing the amount of "looking" you have to do) and is very easy and cheap to care for. The sexes are easy to tell apart, and the female breeds very quickly (up to 100 eggs per day), so you can get your results quickly.

Morgan went on to win the 1933 Nobel Prize in Medicine or Physiology for showing that "genes" are the basis of heredity and are carried on chromosomes. Morgan had originally rejected Darwin's Theory of Sexual Selection – but changed his mind as a direct result of what *D. melanogaster* showed him in the Fly Room. (As an aside, this is the essence of Science – changing your mind when new data comes in.)

It turns out that about three-quarters of human disease genes have a recognisable match in fruit fly DNA. Indeed, the common fruit fly is used to study human conditions such as circadian rhythms, ageing, drug abuse, cancer, diabetes and immunity. It is also an effective genetic model for human neurodegenerative diseases such as Alzheimer's Disease, Huntington's Chorea and Parkinson's Disease.

$23 MILLION FOR FRUIT FLY BOOK?

But even I must admit I was astonished when one of the seminal works in Developmental Biology – Peter Lawrence's book *The Making of a Fly* – was offered on Amazon.com for US$23,698,655.93. Sure, that 1992 book is a classic and it is out of print – so it must be valuable to some people. However, Developmental Biology, while it is fundamental to Evolution, is a fairly narrow field. There aren't many billionaire Developmental Biologists. Overall, $23 million is more than you would reasonably expect to pay.

> The common fruit fly is one of the most widely used organisms in genetic research.

And I was absolutely flabbergasted at how low the shipping price was – only $3.99, for a book costing an enormous $23 million. (Mind you, the author of the book, Peter Lawrence, was absolutely chuffed. At his 70th birthday party he told his guests about how very valued his book was.)

It took a while to solve this mystery. Dr Michael Eisen, an evolutionary biologist at the University of California, Berkeley, worked it out.

A reasonable price range for this excellent, and out of print, book would be $20 to $80 – depending on the condition.

In this case, there were actually two different sellers or retailers offering the book. They had rather obscure names – "profnath" and "bordeebook". They were fairly substantial retailers, with over 8000 (profnath) and 125,000 Seller Ratings (bordeebook) respectively in the previous calendar year. Mind you, one was 16 times bigger than the other.

They were to blame for the inflated price. They each had an Automatic Pricing Program. One program (profnath's) was set to undercut the other by a tiny margin. The other (bordeebook's) was set to charge about one quarter more. These silly programs were sparring with each other inside the dark and noiseless corridors of computer chips – and they were doing this without any human supervision.

Perversely, this example of Free Market Competition did not drive prices down. In fact, the exact opposite happened.

Check it Yourself

Go to Amazon.com. In the "Search" box, click on "All" and select "Books". Click on "Go" at the right side of the "Search" box. The window refreshes, with a slightly different look.

On the next line down, click on "Advanced Search". On the lower right side is a box marked "Sort Results By". By default, it is usually set to "Relevance". Click on Relevance" and select "Price: High to Low". Click on "Search".

Observe and wonder.

The prices are always in the millions – and always absurd.

POWER OF COMPOUND INTEREST

Presumably, Peter Lawrence's fine book had first been offered at around $20 to $80. But then the two Automatic Pricing Programs started, and the power of Compound Interest took over. It took only about 45 days to reach $1.7 million.

When Dr Eisen first checked on 8 April, the smaller retailer (profnath) was charging $1.730 million. It was significantly cheaper than the larger retailer (bordeebook), that offered it for about $468,000 more, at $2.198 million.

The next morning, 9 April, profnath made a huge jump to about $2.194 million – but still managed to undercut the bigger retailer (bordeebook) by the relatively small margin of only $4000. Later that same day, bordeebook responded with a large jump to about $2.788 million.

And the next morning, the pattern repeated. The smaller retailer (profnath) relentlessly followed their larger competitor upward – but always undercutting them by a small amount. And later that day, the bigger retailer (bordeebook) leapfrogged profnath.

	PROFNATH	BORDEEBOOK
8-Apr	$1,730,045.91	$2,198,177.95
9-Apr	$2,194,443.04	$2,788,233.00
10-Apr	$2,783,493.00	$3,536,675.57
11-Apr	$3,530,663.65	$4,486,021.69
12-Apr	$4,478,395.76	$5,690,199.43
13-Apr	$5,680,526.66	$7,217,612.38

UNDERCUT BY 0.17 PER CENT, UP 27 PER CENT, REPEAT

Dr Eisen worked out the pattern.

Early each day, the smaller retailer (profnath) deliberately increased their price to be 99.83 per cent of the bigger retailer's price. They would always undercut them by the same tiny margin.

After a few hours, the larger retailer (bordeebook) would "notice" that their smaller competitor had almost matched them. So bordeebook would then re-price their copy of the book to be about 27 per cent higher than their smaller competitor's offering.

The smaller retailer would then follow the trend upwards. So profnath would offer their copy of the book at 99.83 per cent of bordeebook's new, improved price – always the same percentage (0.17 per cent) cheaper. Then bordeebook would increase their price to be 27 per cent above profnath.

And so the pattern would continue, pushing the book price up and up.

WHY SO CRAZY?

The two book retailers were using Automatic Pricing Programs. These programs didn't have any kind of sanity check.

The two book retailers were using Automatic Pricing Programs. These programs didn't have any kind of sanity check.

The bigger retailer – bordeebook, with 125,000 Seller Ratings – figured that potential buyers would be happy to pay a few dollars more to deal with somebody like them who had a bigger client base. Presumably, the customers of bordeebook figured they got bigger by being more reliable, or quicker to deliver – or something along those lines. Regardless, bordeebook charged an extra 27 per cent (or to be more accurate, 27.0589 per cent more).

The smaller retailer – profnath, with only 8000 Seller Ratings – didn't enjoy such a reputation. So they chose the option of undercutting their larger competitor by a fraction of a per cent – specifically, by 0.17 per cent.

After the price peaked at $23 million, they both dropped their price to around $100. And sure enough, the very next day, the Automatic Pricing Programs started chasing the price up again.

Maybe Not Automatic Pricing Programs?

Maybe I have been tricked. Perhaps these impressive book prices are *not* caused by Automatic Pricing Programs chasing each other forever upwards.

Maybe this is actually some kind of bizarre money-laundering scheme, washing "dirty" money through used books rather than through the more traditional casinos or racetracks? (Is this a new meaning to the phrase "cooking the books"?)

Or perhaps Evil Terrorists were using the prices as a cover for steganography? (That's the process of hiding a secretly coded message in plain view by burying it in something apparently quite innocent.)

But probably not. I tend to follow the old rule, "There's more ignorance and incompetence than malevolence."

So what's driving this process?

Perhaps the computer programs had acquired awareness and consciousness, and had escaped from their human creators? Were they each going through their own internal processes of developmental and evolutionary biology? Probably not . . .

You'd have to think that anyone who fancied this book, and was keeping an eye on the price war, had another option. Once the book had reached $10 million, surely it would have been cheaper to simply approach the author, Peter Lawrence, and offer him $5 million to move in next door and be available to answer your questions 24 hours a day?

These computerised re-pricing programs continue to roam Amazon.com. You would expect law books to be expensive. However, $84 million is a lot, even for a book called *The Law of Mortgages*.

But the record seems to be held by a bioscience book, *Recent Advances in Epilepsy*. It was priced at around $60 trillion – almost four times the US National Debt. Surprisingly, a few weeks later this book on epilepsy had dropped in price to just two cents (yep, $0.02) – with, of course, $3.99 shipping . . .

Value really is in the eye of the beholder.

Pre-owned Book?

On the morning of 26 June 2014, a book written by Professor Hisham Bashir Sharabi (*Arab Intellectuals and the West: The Formative Years, 1875–1914*) was advertised at US$162 million for a new copy.

But a used copy was advertised at $411 million. To justify such a premium price, the previous owner must have been very famous . . .

ACKNOWLEDGMENTS

I deeply thank the person who took out all the boring bits, and left only the good stuff – my Marvelous Mary. Astonishing Alice knocked me out with her punchlines.

As always, this book got off the ground because of all of those on the Publishing side (who would be completely perfect if they were also scientists). So thanks to Claire Craig, Sophie Hamley, Libby Turner, Sarah Fletcher, Emma Rafferty and Charlotte Ree. By working in the name of Books, you are holding a Light against the Darkness of Ignorance.

Thanks also to my University of Sydney work colleagues Caroline Pegram and Shiva Ford, and those at the ABC, Dan Driscoll and David Murray. My gratitude also goes out to those who contributed thoughts, suggestions and ideas via all social media (and any other form of communication so far known, or hereafter to be devised in the foreseeable future forever and ever amen).

Specifically and generally, I thank all the scientists who gave up their time to critique my stories. They include Bryan Burmeister, Siva Shankar, Sheryl van Nunen, Robin Connaughton and Geraint Lewis.

Finally, I thank Nathan Hope for being the first person to use the word "selfie" in print.

REFERENCES

60 Planets to Live On

"Ultimate Solar System Could Contain 60 Earths", by Jacob Aron, *New Scientist*, 30 May 2014, http://www.newscientist.com/article/dn25653-ultimate-solar-system-could-contain-60-earths.html.

"Building the Ultimate Solar System Part 5: Putting the Pieces Together", by Sean Raymond, 23 May 2014, http://planetplanet.net/2014/05/23/building-the-ultimate-solar-system-part-5-putting-the-pieces-together.

Cure Cancer 100 Per Cent?

"Whole Body Irradiation – Radiobiology or Medicine?", by R.H. Mole, *British Journal of Radiology*, May 1953, Vol. 26, No. 305, pages 234–241.

"An Interesting Case of Possible Abscopal Effect in Malignant Melanoma", by D.P.E. Kingsley, *British Journal of Radiology*, October 1975, Vol. 48, No. 574, pages 863–866.

"The Abscopal Effect: Demonstration in Lymphomatous Involvement of Kidneys", by H. Ian Robins et al., *Medical and Pediatric Oncology*, 1981, Vol. 9, No. 5, pages 473–476.

"The Immunobiology of Cancer Immunosurveillance and Immunoediting", by Gavin P. Dunn et al., *Immunity*, August 2004, Vol. 21, No. 2, pages 137–148.

"Regression of Non-Irradiated Metastases After Extracranial Stereotactic Radiotherapy in Metastatic Renal Cell Carcinoma", by Peter J. Wersäll et al., *Acta Oncologica*, January 2006, Vol. 45, No. 4, pages 493–497.

"Immunologic Correlates of the Abscopal Effect in a Patient with Melanoma", by Michael A Prostow et al., *The New England Journal of Medicine*, 8 March 2012, Vol. 366, No. 10, pages 925–931.

"Abscopal Effects After Conventional and Stereotactic Lung Irradiation of Non-Small-Cell Lung Cancer", by Shankar Siva et al., *Journal of Thoracic Oncology*, August 2013, Vol. 8, No. 8, pages e71-e72.

"Disappearing Act", by James Welsh, *Discover*, March 2014, pages 24–26.

"Abscopal Effects of Radiation Therapy: A Clinical Review for the Radiobiologist", by Shankar Siva et al., *Cancer Letters*, Article in Press, published online 11 October 2013, pages 1–9.

"Beating the Big C", *New Scientist*, 23 November 2013, Vol. 220, No. 2944, page 3.

"Cancer Meets Its Nemesis", by Andy Coghlan, *New Scientist*, Vol. 220, No. 2944, 23 November 2013, pages 6–7.

Ahead of His Time

"Man Hears People Speak Before Seeing Lips Move", by Helen Thomson, *New Scientist*, 6 July 2013, Vol. 219, No. 2924, page 11.

"Sight and Sound Out of Synch: Fragmentation and Renormalisation of Audiovisual Integration and Subjective Timing" by Elliot D. Freeman et al., *Cortex*, November–December 2013, Vol. 49, No. 10, pages 2875–2887.

Apple Logo and Alan Turing

"Interview with Rob Janoff, Designer of the Apple Logo", by Ivan Raszl, 3 August 2009, http://creativebits.org/interview/interview_rob_janoff_designer_apple_logo.

"Steve Jobs and the Evolution of the Apple Logo: 'Don't Make It Cute'", by Abe Sauer, 9 October 2011, http://www.brandchannel.com/home/post/steve-jobs-evolution-apple-logo.aspx.

"At Bletchley Park, a Reminder About the History of Cracking Codes", by J.D. Biersdorfer, *The New York Times Bits* blog, 9 August 2014, http://bits.blogs.nytimes.com/2014/08/09/at-bletchley-park-a-reminder-about-the-history-of-cracking-codes.

Arctic Meltdown – Milankovitch Cycles

"Sharply Increased Mass Loss from Glaciers and Ice Caps in the Canadian Arctic Archipelago", by Alex S. Gardner et al., *Nature*, 19 May 2011, Vol. 473, No. 7347, pages 357–360.

"Reconstructed Changes in the Arctic Sea Ice Over the Past 1,450 Years", by Christophe Kinnard et al., *Nature*, 24 November 2011, Vol. 479, No. 7374, pages 509–512.

"Arctic Sea Ice: What, Why and What Next", by Ramez Naam, *Scientific American* Guest Blog, 21 September 2012, http://blogs.scientificamerican.com/guest-blog/2012/09/21/arctic-sea-ice-what-why-and-what-next.

"The Great Thaw", by Anil Ananthaswamy, *New Scientist*, Vol. 216, No. 2889, 3 November 2012, pages 32–35.

"Meltwater Routing and the Younger Dryas", by Alan Condron and Peter Winsor, *PNAS*, Vol. 109, No. 49, 4 December 2012, pages 19928–19933.

"Global Warning", by Michael LePage, *New Scientist*, 17 November 2012, Vol. 216, No. 2891, pages 34–39.

Bank Robbery

"Crime and Punishment: An Economic Approach", by Gary S. Becker, *Journal of Political Economy*, March–April 1968, Vol. 76, No. 2, pages 169–217.

Freakonomics: A Rogue Economist Explores the Hidden Side of Everything, by Steven Levitt and Stephen J. Dubner, William Morrow, New York, 2005.

"'Dumb and Dumber' Bank Robber Completes US Jail Term", by Peter Mitchell, *Herald Sun*, 1 March 2009.

"Dumb and Dumber", *Australian Story*, ABC TV, program transcript, 12 April 2010.

"Robbing Banks: Crime Does Pay – But Not Very Much", by Barry Reilly et al., *Significance*, June 2012, Vol. 9, No. 3, pages 17–21.

"Gary Becker: Real-World Economist", by Lawrence H. Summers, *Time*, 19 May 2014, page 13.

Calorie is Not a Calorie

"The Last Word: A Balanced Diet", *New Scientist* (Australian edition), 22 May 2010, Vol. 206, No. 2761, page 57.

"Postprandial Energy Expenditure in Whole-Food and Processed-Food Meals: Implications for Daily Energy Expenditure", by Sadie B. Barr and Jonathan C. Wright, *Food and Nutrition Research*, 2 July 2010, Vol. 54, pages 1-10.

"Discrepancy between the Atwater Factor Predicted and Empirically Measured Energy Values of Almonds in Human Diets", by Janet A. Novotny et al., *American Journal of Clinical Nutrition*, August 2012, Vol. 96, No. 2, pages 296–301.

"Everything You Know about Calories is Wrong", by Rob Dunn, *Scientific American* (Australian edition), September 2013, Vol. 309, No. 3, pages 46-49.

"Association between Chocolate Consumption and Fatness in European Adolescents", by Magdalena Cuenca-Garcia et al., *Nutrition*, February 2014, Vol. 30, No. 2, pages 236–239.

"'Fed Up' Asks, Are All Calories Equal?", by Anahad O'Connor, *The New York Times* Well blog, 9 May 2014, http://well.blogs.nytimes.com/2014/05/09/fed-up-asks-are-all-calories-equal.

Cigarettes: Roll Your Own Versus Factory Made

"Hand-Rolled Cigarette Smoking and Risk of the Cancer of the Mouth, Pharynx and Larynx", by Eduardo De Stefani et al., *Cancer*, 1 August 1992, Vol. 70, No. 3, pages 679–682.

"Could Czech Smokers Help Balance Budget? Philip Morris Offers Dubious Cost/Benefit Data", by Gordon Fairclough, The Wall Street Journal Europe, 16 July 2001.

"The Costs of Tobacco, Alcohol and Illicit Drug Abuse to Australian Society in 2004/05", by David J. Collins and Helen M. Lapsley, Department of Health and Ageing, Canberra, 2008, http://www.health.gov.au/internet/drugstrategy/publishing.nsf/Content/34F55AF632F67B70CA2573F60005D42B/$File/mono64.pdf.

"Prevalence and Correlates of Roll-Your-Own Smoking in Thailand and Malaysia: Findings of the ITC-South East Asia Survey", by David Young et al., *Nicotine and Tobacco Research*, May 2008, Vol. 10, No. 5, pages 907–915.

"The Importance of Tobacco Prices to Roll-Your-Own (RYO) Smokers (National Survey Data): Higher Tax Needed on RYO", *The New Zealand Medical Journal*, Vol. 122, No. 1305, 30 October 2009, pages 92–96.

"Merchants of Doubt: How a Handful of Scientists Obscured the Truth on Issues from Tobacco Smoke to Global Warming", by Naomi Oreskes and Erik M. Conway, Bloomsbury Press, New York, 2010.

"Prevalence, Correlates of, and Reasons for Using Roll-Your-Own Tobacco in a High RYO Use Country: Findings from the ITC New Zealand Survey", by David Young et al., *Nicotine and Tobacco Research*, November 2011, Vol. 12, No. 11, pages 1089–1098.

"Trends in Roll-Your-Own Smoking: Findings from the ITC Four-Country Survey (2002–2008)", by David Young et al., *Journal of Environmental and Public Health*, 12 May 2012, Vol. 2012, pages 1–7.

"Nicotine-, Tobacco Particulate Matter- and Methamphetamine-Produced Locomotor Sensitisation in Rats", by Katherine A. Brennan et al., *Psychopharmacology*, 4 August 2013, Vol. 228, No. 4, pages 659–672.

"Roll Your Own Cigarettes Are Less Natural and At Least As Harmful As Factory Rolled Tobacco", by Richard Edwards, *The British Medical Journal*, 11 February 2014, Vol. 348, No. f7616.

Dancing with Deer Evolves Your Brain

"Tracking Down the Deer Runner", by Michael J. Joyner, *The New York Times*, 16 May 2014. http://well.blogs.nytimes.com/2014/05/16/tracking-down-the-deer-runner/.

"Endurance Running and the Evolution of *Homo*", by Dennis M. Bramble and Daniel E. Lieberman, *Nature*, 18 November 2004, Vol. 432, No. 7015, pages 345–352.

"Rise of the Human Predator", by Kate Wong, *Scientific American*, April 2014, Vol. 310, No. 4, pages 32–37.

Eat Less, Move More

"Comparison of Weight-Loss Diets with Different Compositions of Fat, Protein, and Carbohydrates", by Frank M. Sacks et al., *New England Journal of Medicine*, 26 February 2009, Vol. 360, No. 9, pages 859–873.

"Calorie Counters", editorial, *The New York Times*, 3 February 2010.

The Five-Second Rule

"Residence Time and Food Contact Time Effects on Transfer of Salmonella Typhimurium from Tile, Wood and Carpet: Testing the Five-Second Rule", by P. Dawson et al., *Journal of Applied Microbiology*, April 2007, Vol. 102, No. 4, pages 945–953.

"The Five-Second Rule Explored, or How Dirty Is That Bologna", by Harold McGee, *The New York Times*, 9 May 2007.

"The 5-Second Rule", by C. Claiborne Ray, *The New York Times*, 28 February 2011.

"Fact or Fiction?: The 5-Second Rule for Dropped Food", by Larry Greenemeier, *Scientific American*, 25 March 2014, http://www.scientificamerican.com/article/fact-or-fiction-the-5-second-rule-for-dropped-food.

Frog Milkshake

"Composition and Antimicrobial Activity of the Skin Peptidome of Russian Brown Frog *Rana temporaria*", by T.Y. Samgina et al., *Journal of Proteome Research*, December 2012, Vol. 11, No. 12, pages 6213–6222.

"Milking Frog Skin", by Jennifer Abbasi, *Discover*, May 2014, Vol. 35, No. 4, page 22.

"A Brief History of Ice", by Amanda Green, *Popular Mechanics*, June 2014, page 124.

Gold in Trees

"Bacteria Help Grow Gold Nuggets from Dirt", by Richard A. Kerr, *Science*, 14 July 2006, Vol. 313, No. 5784, page 159.

"Biomineralization of Gold: Biofilms on Bacterioform Gold", by Frank Reith et al., *Science*, 14 July 2006, Vol. 313, No. 5784, pages 233–236.

"Tip for Gold Prospectors: Follow the Bacteria", by Henry Fountain, *The New York Times*, 18 July 2006.

"Accretion of the Earth and Segregation of Its Core", by Bernard J. Wood et al., *Nature*, 15 June 2006, Vol. 441, No. 7095, pages 825–833.

"Gold Biomineralization by a Metalophore from a Gold-Associated Microbe", by Chad W. Johnston et al., *Nature Chemical Biology*, April 2013, Vol. 9, No. 4, pages 241–243.

"Flash Vaporization During Earthquakes Evidenced by Gold Deposits", by Dion K. Weatherley and Richard W. Henley, *Nature Geoscience*, April 2013, Vol. 6, No. 4, pages 294–298.

"Gold Treasure Marked Out by Tree Leaves", by Michael Slezak, *New Scientist*, Vol. 220, No. 2940, 26 October 2013, page 19.

"Natural Gold Particles in *Eucalyptus* Leaves and Their Relevance to Exploration for Buried Gold Deposits", by Melvyn Lintern et al., *Nature Communications*, 22 October 2013, pages 1–7.

"A Nugget of Truth Way Out West", by Michael West, *The Sydney Morning Herald*, Weekend Business, 26–27 October 2013.

Greed is not Good

"Social Class, Contextualism, and Empathic Accuracy", by Michael W. Kraus, *Psychological Science*, November 2010, Vol. 21, No. 11, pages 1716–1723.

"What Price Will We Pay for Greed: Why Greed May Be Our Undoing", by Ray B Williams, *Psychology Today* Wired for Success blog, 29 February 2012, http://www.psychologytoday.com/blog/wired-success/201202/what-price-will-we-pay-greed.

"Greed on Wall Street Prevents Good from Happening", by Dacher Keltner and Paul Piff, *The New York Times*, 16 March 2012.

"Higher Social Class Predicts Increased Unethical Behaviour", by Paul Piff et al., *PNAS*, 13 March 2012, Vol. 109, No. 11, pages 4086–4091.

"Evidence that Publication Bias Contaminated Studies Relating Social Class and Ethical Behaviour", by Gregory Francis, *PNAS*, 19 June 2012, Vol. 109, No. 25, E1587.

"Reply to Francis: Cumulative Power Calculations Are Faulty when Based on Observed Power and a Small Sample Of Studies", by Paul Piff et al., *PNAS*, 19 June 2012, Vol. 109, No. 25, E1588.

"More Bounty, No Heart: Our Selfish Rich", by Julia Baird, *The Sydney Morning Herald*, News Review, October 12–13, 2013, page 12.

"Wealth and the Inflated Self: Class, Entitlement, and Narcissism", by Paul Piff, *Personality and Social Psychology Bulletin*, January 2014, Vol. 40, No. 1, pages 34–43.

"Inequality Is a Drag", by Paul Krugman, *The New York Times*, 7 August 2014.

Green Tea – Leaded or Unleaded?

"Scale and Causes of Lead Contamination in Chinese Tea", by Wen-Yan Han et al., *Environmental Pollution*, January 2006, Vol. 139, No. 1, pages 125–132.

"Hepatotoxicity Associated with Supplements Containing Chinese Green Tea (*Camellia sinensis*)", by Herbert L. Bonkovsky, *Annals of Internal Medicine*, 3 January 2006, Vol. 144, No. 1, pages 68–71.

"Beneficial Effects of Green Tea – A Review", by Carmen Cabrera et al., *Journal of the American College of Nutrition*, April 2006, Vol. 25, No. 2, pages 79–99.

"Possible Controversy over Dietary Polyphenols: Benefits vs Risks", by Joshua D. Lambert et al., *Chemical Research in Toxicology*, April 2007, Vol. 20, No. 4, pages 583–585.

"What's in Your Green Tea?", by Anahad O'Connor, *The New York Times* Well blog, 23 May 2013, http://well.blogs.nytimes.com/2013/05/23/whats-in-your-green-tea/.

"Product Review: Green Tea Supplements, Drinks and Brewable Teas Review", by ConsumerLab.com, initially posted 21 December 2012, updated 20 June 2014.

Lactose Intolerance

"Lactose Tolerance in East Africa Points to Recent Evolution", by Nicholas Wade, *The New York Times*, 11 December 2006.

"There's More Than One Way to Have Your Milk and Drink It, Too", by Ann Gibbons, *Science*, 15 December 2006, Vol. 314, No. 5806, page 1672.

"How Africa Learned to Love The Cow", by Erika Check, *Nature*, 21 December 2006, Vol. 444, No. 7122, pages 994–996.

"The Origins of Lactase Persistence in Europe", by Yuval Itan et al., *PLoS Computational Biology*, August 2009, Vol. 5, No. 8, e1000491, pages 1–13.

"Evolution of Lactase Persistence: An Example of Human Niche Construction", by Pascale Gerbault et al., *Philosophical Transactions of the Royal Society B*, 27 March 2011, Vol. 366, No. 1566, pages 863–877.

"The Evolution of Lactase Persistence in Europe. A Synthesis of Archaeological and Genetic Evidence", by Michela Leonardi et al., *International Dairy Journal*, February 2012, Vol. 22, No. 2, pages 88–97.

"Dietary and Biological Factors Influencing Lactose Intolerance", by O. Brown-Esters et al., *International Dairy Journal*, February 2012, Vol. 22, No. 2, pages 98–103.

"Earliest Evidence for Cheese Making in the Sixth Millennium BC in Northern Europe", by Mélanie Salque et al., *Nature*, 24 January 2013, Vol. 493, No. 7433, pages 522–525.

"Archaeology: The Milk Revolution", by Andrew Curry, *Nature*, 1 August 2013, Vol. 500, No. 7460, pages 20–22.

Meat Allergy

"One Tick Red Meat Can Do Without", by Bianca Nogrady, *The Australian*, 10 May 2008.

"Tick Bite, Red Meat Allergy Link 'Found'", by Danny Rose, *The Sydney Morning Herald*, 4 May 2009.

"An Association between Tick Bite Reactions and Red Meat Allergy in Humans", by Sheryl A. van Nunen et al., *The Medical Journal of Australia*, 4 May 2009, Vol. 190, No. 9, pages 510–511.

"Identification of Galactose-a-1,3-Galactose in the Gastrointestinal Tract of the Tick *Ixodes ricinus*: Possible Relation with Red Meat Allergy", by C. Hamsten et al., *Allergy*, April 2013, Vol. 68, No. 4, pages 549–552.

"Tick Bite Anaphylaxis: Incidence and Management in an Australian Emergency Department", by Tristan B. Rappo et al., *Emergency Medicine Australasia*, August 2013, Vol. 25, No. 4, pages 297–301.

"Just One Bite: Ticks and Allergies on the North Shore", by Louise Williams, *The Sydney Morning Herald*, 27 December 2013.

Nazis Stole Space Buddha

"The Use of Meteoric Iron", by T.A. Rickard, *The Journal of the Royal Anthropological Institute of Great Britain and Ireland*, 1941, Vol. 71, No. 1/2, pages 55–66.

"Meteors and Meteorites in the Ancient Near East", by Judith Kingston Bjorkman, *Meteoritics*, June 1973, Vol. 8, No. 2, pages 91–132.

"On the Use of Iron by the Eskimos in Greenland", by Vagn Fabritius Buchwald, *Materials Characterization*, September 1992, Vol. 29, No. 2, pages 139–176.

"Buddha from Space – An Ancient Object of Art Made of a Chinga Iron Meteorite Fragment", by Elmar Buchner et al., *Meteoritics & Planetary Science*, September 2012, Vol. 47, No. 9, pages 1491–1501.

"5,000 years old Egyptian Iron Beads Made from Hammered Meteoritic Iron", by Thilo Rehren et al., *Journal of Archaeological Research*, 31 July 2013, Vol. 40, pages 4785–4792.`

Orgasms via Foot

"Foot Orgasm Syndrome: A Case Report in a Woman", by Marcel D. Waldinger et al., *Journal of Sexual Medicine*, August 2013, Vol. 10, No. 8, pages 1926–1934.

Paleolithic Diet

"Paleolithic Nutrition – A Consideration of Its Nature and Current Implications", by S. Boyd Eaton and Melvin Konner, *The New England Journal of Medicine*, 31 January 1985, Vol. 312, No. 5, pages 283–289.

"How to Run on Thin Air" by Nicola Jones, *New Scientist*, Vol. 176, No. 2373, 14 December 2002.

"Thirty Thousand-Year-Old Evidence of Plant Food Processing", by Anna Revedin et al., *PNAS*, 2 November 2010, Vol. 107, No. 44, pages 18815–18819.

"Neanderthal Use of Fish, Mammals, Birds, Starchy Plants and Wood 125–250,000 Years Ago", by Bruce L. Hardy and Marie-Hélène Moncel, *PLoS ONE*, August 2011, Vol. 6, No. 8, e23768, pages 1–9.

"Neanderthals' Love of Fine Dining", *New Scientist*, 29 October 2011, Vol. 211, No. 2836, page 70.

"Human Ancestors Were Nearly All Vegetarians", by Rob Dunn, *Scientific American* Guest Blog, 23 July 2012, http://blogs.scientificamerican.com/guest-blog/2012/07/23/human-ancestors-were-nearly-all-vegetarians.

Paleofantasy: What Evolution Really Tells Us about Sex, Diet, and How We Live, by Marlene Zuk, W. W. Norton & Company, New York, 2013.

"Should We Aim to Live Like Cavemen?", by Alison George, *New Scientist*, 29 March 2013, Vol. 217, No. 2909, pp28–29.

"Impacts of Plant-Based Foods in Ancestral Hominin Diets on the Metabolism and Function of Gut Microbiota *in Vitro*", by Gary S. Frost et al., *mBio*, 20 May 2014, Vol. 5, No. 3, e00853–14, pages 1–11.

"How to Really Eat Like a Hunter Gatherer: Why the Paleo Diet Is Half-Baked", by Ferris Jabr, *Scientific American*, 3 June 2013, http://www.scientificamerican.com/article/why-paleo-diet-half-baked-how-hunter-gatherer-really-eat/.

"Health Myths: We Should Live and Eat Like Cavemen", by Caroline Williams, *New Scientist*, 24 August 2013, Vol. 219, No. 2931, page 36.

"Health News Examined", *Time*, 16 June 2014, page 13.

"Dental Calculus Reveals Unique Insights into Food Items, Cooking and Plant Processing in Prehistoric Central Sudan", by Stephen Buckley et al., *PLoS One*, July 2014, Vol. 9, No. 7, e100808, pages 1–10.

Permafrost Feedback Loop

"Climate Change: High Risk of Permafrost Thaw", by Edward A.G. Schuur and Benjamin Abbott, *Nature*, 1 December 2011, Vol. 480, No. 7375, pages 32–33.

"As Permafrost Thaws, Scientists Study the Risks", by Justin Gillis, *The New York Times*, 16 December 2011.

"Policy Implications of Warming Permafrost", United Nations Environment Program, Nairobi, 2012, http://www.unep.org/pdf/permafrost.pdf.

"Speleothems Reveal 500,000-Year History of Siberian Permafrost", by A. Vaks et al., *Science Express*, 21 February 2013, pages 1–6.

"Vast Costs of Arctic Change" by Gail Whiteman et al., *Nature*, 25 July 2013, Vol. 499, No. 7459, pages 401–403.

"Ebullition and Storm-Induced Methane Release from the East Siberian Arctic Shelf", by Natalia Shakhova et al., *Nature Geoscience*, 24 November 2013, Vol. 7, No. 1, pages 64–70.

Selfie

"Mandela Funeral Selfie Adds to Image Problem for Denmark's Prime Minister", by Andrew Anthony, *The Observer*, 15 December 2013.

"Art at Arm's Length: A History of the Selfie", by Jerry Saltz, *New York Magazine*, 3 February 2014.

"Celeb Rover Marks One Martian Year on Mars with Selfie", by Jacob Aron, *New Scientist*, 24 June 2014, http://www.newscientist.com/article/dn25779-celeb-rover-marks-one-martian-year-on-mars-with-selfie.html#.U_v2c0g7bB8.

"When Selfies Won't Do", by Alex Williams, *The New York Times*, 10 July 2014.

"Scrabble Approves Selfie, Te and Bromance for Dictionary", *The Guardian*, 5 August 2014.

Sleep, Mysterious Sleep

"Unconscious Networking", by Mark A. Pinsk and Sabine Kastner, *Nature*, 3 May 2007, Vol. 447, No. 7140, pages 46–47.

"Intrinsic Functional Architecture in the Anaesthetized Monkey Brain", by J.L. Vincent et al., *Nature*, 3 May 2007, Vol. 447, No. 7140, pages 83–86.

"Why Sleep?", by Richard Gallagher, *The Scientist*, April 2009, Vol. 23, No. 4, page 15.

"Disappearing Before Dawn", by Kelly Rae Chi, *The Scientist*, April 2009, Vol. 23, No. 4, pages 34–40.

"The Gears of the Sleep Clock", by Allan Pack, *The Scientist*, April 2009, Vol. 23, No. 4, pages 43–47.

"Inducing Sleep by Remote Control Facilitates Memory Consolidation in *Drosophila*", by Jeffrey M. Donlea et al., *Science*, 24 June 2011, Vol. 332, No. 6037, pages 1571–1576.

"Sleep and Synaptic Homeostasis: Structural Evidence in *Drosophila*", by Daniel Bushey et al., *Science*, 24 June 2011, Vol. 332, No. 6037, pages 1576–1581.

"Synaptic Plasticity by Antidromic Firing During Hippocampal Network Oscillations", by Olena Bukalo et al., *PNAS*, 26 March 2013, Vol. 110, No. 13, pages 5175–5180.

"Sleep: The Brain's Housekeeper?" by Emily Underwood, *Science*, 18 October 2013, Vol. 342, No. 6156, page 301.

"Sleep It Out", by Suzana Herculano-Houzel, *Science*, 18 October 2013, Vol. 342, No. 6156, pages 316–317.

"Sleep Drives Metabolite Clearance from the Human Brain", by Lulu Xie et al, *Science*, 18 October 2013, Vol. 342, No. 6156, pages 373–377.

Smarter than Your Parents, Not as Smart as Your Kids

"Flynn's Effect: Intelligence Scores Are Rising, James R. Flynn Discovered – But He Remains Very Sure We're Not Getting Any Smarter", by Marguerite Holloway, *Scientific American*, January 1999, Vol. 280, No. 1, pages 37–38.

"Solving the IQ Puzzle", by James R. Flynn, *Scientific American Mind*, October/November 2007, Vol. 18, No. 5, pages 24–31.

"None of the Above: What IQ Doesn't Tell You About Race", by Malcolm Gladwell, *The New Yorker*, 17 December 2007.

"Parasite Prevalence and the Worldwide Distribution of Cognitive Ability", by Christopher Eppig et al., *Proceedings of the Royal Society B*, 22 December 2010, Vol. 277, No. 1701, pages 3801–3808.

"Intelligence: New Findings and Theoretical Developments", by Richard E. Nisbett et al., *American Psychologist*, February–March 2012, Vol. 67, No. 2, pages 130–159.

"The Flynn Effect in Korea: Large Gains", by Jan te Nijenhuis et al., *Personality and Individual Differences*, July 2012, Vol. 53, No. 2, pages 147–151.

"A Life History Model of the Lynn-Flynn Effect", by Michael A. Woodley, *Personality and Individual Differences*, July 2012, Vol. 53, No. 2, pages 152–156.

"Richard Lynn's Contributions to Personality and Intelligence", by James Thompson, *Personality and Individual Differences*, July 2012, Vol. 53, No. 2, pages 157–161.

Are We Getting Smarter? Rising IQ in the 21st Century, by James R. Flynn, Cambridge University Press, Cambridge, 2012.

"Can We Keep Getting Smarter?", by Tim Folger, *Scientific American*, September 2012, Vol. 307, No. 3, pages 30–33.

"*Homo* (Sans) *Sapiens*: Is Dumb and Dumber Our Evolutionary Destiny?", by Gary Stix, *Scientific American* Blog Network, 26 November 2012, http://blogs.scientificamerican.com/talking-back/2012/11/26/homo-sans-sapiens-is-dumb-and-dumber-our-evolutionary-destiny.

"The Cognitive Effects of Micronutrient Deficiency: Evidence From Salt Iodization in the United States", by James Feyrer et al., The National Bureau Of Economic Research, Working Paper No. 19233, July 2013.

"Gauging the Intelligence of Infants", by Kenneth Chang, *The New York Times*, 7 April 2014.

"Chimpanzee Intelligence Is Heritable", by William D. Hopkins et al., *Current Biology*, 21 July 2014, Vol. 24, No. 14, pages 1649–1652.

Sneeze in Sunlight

"Reflexe Photo-Sternutatoire [Photosternutatory Reflex]", by J. Sedan, *Revue d'Oto-Neuro-Ophtalmologie*, 1954, Volume 26, No. 2, pages 123–126.

"Sneezing in Response to Light", by H.C. Everett, *Neurology*, May 1964, Volume 14, No. 5, pages 483–490.

"ACHOO Syndrome (Autosomal Dominant Compelling Helio-Ophthalmic Outburst Syndrome)", by W.R. Collie et al., *Birth Defects Original Article Series*, 1978, Vol. 14, No. 6B, pages 361–363.

"The Photic Sneeze Reflex as a Risk Factor to Combat Pilots", by R.A. Breitenbach et al., *Military Medicine*, December 1993, Vol. 158, No. 12, pages 806–809.

"Sneezing Reflex Associated with Intravenous Sedation and Periocular Anesthetic Injection" by Eric S. Ahn et al., *American Journal of Ophthalmology*, July 2009, Vol. 146, No. 1, pages 31–35.

"Sneeze Reflex: Facts and Fiction", by Murat Songu and Cernal Cingi, *Therapeutic Advances in Respiratory Diseases*, June 2009, Vol. 3, No. 3, pages 131–141.

"When the Sun Prickles Your Nose: An EEG Study Identifying Neural Bases of Photic Sneezing", by Nicolas Langer et al., *PLoS One*, February 2010, Vol. 5, No. 2, e9208, pages 1–6.

"Web-Based, Participant-Driven Studies Yield Novel Genetic Associations for Common Traits", by Nicholas Eriksson et al., *PLoS Genetics*, 24 June 2010, Vol. 6, No. 6, e1000993, pages 1–20.

"Violent Expiratory Events: On Coughing and Sneezing", by Lydia Bourouiba et al., *Journal of Fluid Mechanics*, March 2014, Vol. 745, pages 537–563.

Space, The Hostile Frontier (More than a Bad Hair Day)

"Astronaut Grows Too Tall", *The New York Times*, 12 July 1994.

"Orbital and Intracranial Effects of Microgravity: Findings at 3-T MR Imaging", by Larry A. Kramer et al., *Radiology*, June 2012, Vol. 263, No. 3, pages 819–827.

"Exercise in Space: Human Skeletal Muscle after 6 Months Aboard the International Space Station", by Scott Trappe et al., *Journal of Applied Physiology*, 1 April 2009, Vol. 106, No. 4, pages 1159–1168.

"Spaceflight Promotes Biofilm Formation by *Pseudomonas aeruginosa*", by Wooseong Kim et al., *PLoS One*, April 2013, Vol. 8, No. 4, e62437.

"Finding New Apps", by Frank Morring, Jr, *Aviation Week & Space Technology*, 28 April 2014, Vol. 176, No. 14, page 19.

Swearing and Pain

"Is Spirituality a Critical Ingredient of Meditation? Comparing the Effects of Spiritual Meditation, Secular Meditation, and Relaxation on Spiritual, Psychological, Cardiac, and Pain Outcomes", by Amy B. Wachholtz and Kenneth I. Pargament, *Journal of Behavioural Medicine*, August 2005, Vol. 28, No. 4, pages 369–384.

"Almost Before We Spoke, We Swore", by Natalie Angier, *The New York Times*, 20 September 2005.

"Swearing as a Response to Pain", by Richard Stephens et al., *NeuroReport*, 5 August 2009, Vol. 20, No. 12, pages 1056–1060.

"Why The #$%! Do We Swear? For Pain Relief", by Frederik Joelving, *Scientific*, 12 July 2009, http://app1.scientificamerican.com.

"Swearing Can Help Relieve Pain, Study Claims", *The Telegraph*, 18 April 2011.

"Swearing As A Response To Pain – Effect Of Daily Swearing Frequency", by Richard Stephens and Claudia Umland, *The Journal of Pain*, December 2011, Vol. 12, No. 12, pages 1274–1281.

Target Knows Before It Shows

"Predicting the Quality and Prices of Bordeaux Wine", by Orley Ashenfelter, *The Economic Journal*, June 2008, Vol. 118, No. 529, pages F174–F184.

Race Against the Machine: How the Digital Revolution Is Accelerating Innovation, Driving Productivity, and Irreversibly Transforming Employment and the Economy, by Erik Brynjolfsson and Andrew McAfee, Digital Frontier Press, 2011.

"How Companies Learn Your Secrets", by Charles Duhigg, *The New York Times*, 16 February 2012.

"Predicting Consumer Behaviour with Web Search", by Sharad Goel et al., *PNAS*, 12 October 2012, Vol. 107, No. 41, pages 17486–17490.

"Will a Robot Take Your Job?", by Gary Marcus, *The New Yorker News Desk* blog, 29 December 2012, http://www.newyorker.com/online/blogs/newsdesk/2012/12/will-robots-take-over-our-economy.html.

"The Parable of Google Flu: Traps in Big Data Analysis", by David Lazer et al., *Science*, 14 March 2014, Vol. 343, No. 6176, pages 1203–1205.

"Wikipedia Usage Estimates Prevalence of Influenza-Like Illness in the United States in Near Real Time", by David J. McIver and John S. Brownstein, *PLoS Computational Biology*, April 2014, Vol. 10, No. 4, e1003581, pages 1–8.

"Love, Actually. Algorithms: Data Analyses Beat the Educated Guess Almost Every Time", by Tom Chivers, *The Australian Financial Review*, 28 February 2014, page 8R.

"Big Data", by Andrew McAfee and Erik Brynjolfsson, *Harvard Business Review*, 18 March 2014, pages 60–67.

"Out in the Cold", by Aaron Patrick, *The Australian Financial Review*, 8 May 2014, pages 52–53.

"Who Watches the Watchers? Big Data Goes Unchecked", by Josh Gerstein and Stephanie Simon, *Politico*, 14 May 2014, http://www.politico.com/story/2014/05/big-data-beyond-the-nsa-106653.html.

The World's Most Expensive Book

"Feedback: Biology Book Only Costs $23 Million", *New Scientist*, 29 March 2013, Vol. 217, No. 2909, page 264.

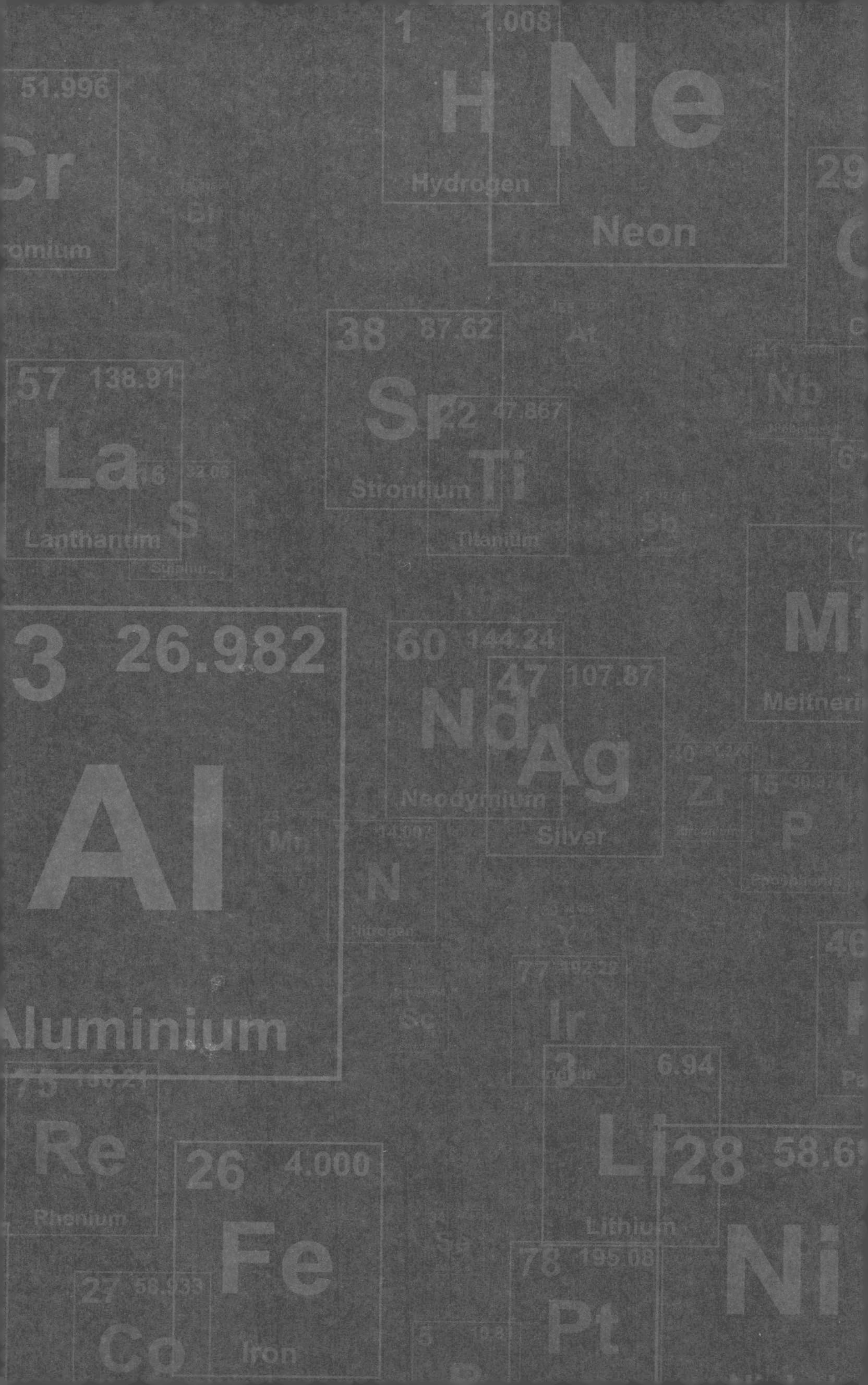

1
1.008
H
Hydrogen
Ne
Neon
51.996
Cr
38
87.62
Sr
Strontium
57
138.91
La
Lanthanum
22
47.867
Ti
S
3
26.982
Al
Aluminium
60
144.24
Nd
Neodymium
47
107.87
Ag
Silver
N
Ir
3
6.94
Li
Lithium
Re
Rhenium
26
Fe
Iron
28
Ni
78
Pt
27
Co